Two Manuals Combined into One Book

Combat and Operational Stress Control

And

Marine Combat Stress

Department of Defense

Pentagon
Publishing

Field Manual
No. 4-02.51 (8-51)

Headquarters
Department of the Army
Washington, DC, 6 July 2006

COMBAT AND OPERATIONAL STRESS CONTROL

Contents

Distribution Restriction: Approved for public release; distribution is unlimited.

***This publication supersedes FM 8-51, 29 September 1994.**

Figures

Tables

Preface

This publication outlines the functions and operations of each combat and operational stress control (COSC) element within an area of operations (AO). This field manual (FM) establishes Army doctrine and provides guidance for conducting COSC support for combat, stability, and reconstruction operations from brigade to theater level. The information provided in this publication will assist commanders and their staffs to operate efficiently at all levels of command and throughout the operational continuum. It may be used by medical planners to supplement FM 4-02, FM 8-42 (4-02.42), and FM 8-55 (4-02.55). Users of this publication should be familiar with Army Regulation (AR) 40-216 and with FM 4-02, FM 4-02.6, FM 4-02.10, FM 4-02.21, FM 4-02.24, FM 6-22.5, FM 8-10-6, FM 8-10-14, and FM 22-51.

This manual is in consonance with FM 7-15, Army Universal Task List (AUTL) and support the following Army tactical task (ART) provided below. Commanders should use the AUTL as a cross-reference for tactical tasks. The AUTL provides a standard doctrinal foundation and catalogue of the Army's tactical collective tasks.

AUTL ART	
ART 6.5.1	Provide Combat Casualty Care
ART 6.5.1.5	Provide Mental Health/Neuropsychiatric Treatment
ART 6.5.4	Provide Casualty Prevention
ART 6.5.4.5	Provide Combat Operational Stress Control Prevention

The staffing and organization structure presented in this publication reflects those established in the base tables of organization and equipment (base TOE) and are current as of the publication print date. Such staffing is subject to change to comply with manpower requirements criteria outlined in AR 71-32. Those requirements criteria are also subject to change if the modification table of organization and equipment (MTOE) is significantly altered.

Users of this publication are encouraged to submit comments and recommendations to improve the publication. Comments should include the page, paragraph, and line(s) of the text where the change is recommended. The proponent of this publication is the United States (US) Army Medical Department Center and School (USAMEDDC&S). Comments and recommendations should be forwarded in a letter format directly to the **Commander, USAMEDDC&S, ATTN: MCCS-FCD-L, 1400 East Grayson Street, Fort Sam Houston, Texas 78234-5052**, or at e-mail address: Medicaldoctrine@amedd.army.mil.

This FM applies to the Active Army, the Army National Guard (ARNG)/Army National Guard of the United States (ARNGUS), and the United States Army Reserve (USAR) unless otherwise stated.

Unless this publication states otherwise, masculine nouns and pronouns do not refer exclusively to men.

Use of trade or brand names in this publication is for illustrative purposes only and does not imply endorsement by the Department of Defense (DOD). The Army Medical Department (AMEDD) is in a transitional phase with terminology. This manual uses the most current terminology; however, other FM 4-02-series and FM 8-series may use the older terminology. Changes in terminology are a result of adopting the terminology currently used in the joint and/or North Atlantic Treaty Organization (NATO) and American, British, Canadian, and Australian (ABCA) Armies publication arenas. Therefore, the following terms are synonymous and the current terms are listed first, to include—

- Force health protection (FHP), health service support (HSS), and combat health support (CHS).
- Medical logistics (MEDLOG), health service logistics (HSL), and combat health logistics (CHL).
- Levels of care, echelons of care, and roles of care.

- Combat and operational stress control (COSC) and combat stress control (CSC).
- Behavioral health (BH) and mental health (MH).

Additionally, please note that the term "battle fatigue (BF)" that is used in AR 40-216 is being replaced with the term "combat stress reaction (CSR)." This change is due to DOD Directive (DODD) 6490.5, which specified that all Military Services use the term CSR for the purpose of joint interoperability. A DOD (Health Affairs) working group with the Services later added the term "operational stress reaction (OSR)" to further characterize stress reactions experienced by Soldiers. The only difference between a CSR and an OSR is that it takes place in a combat environment. To reduce confusion from this change in terminology, this manual will use combat and operational stress reaction (COSR) as opposed to "BF". The COSR (battle fatigue) "casualties" are Soldiers experiencing a stress reaction in combat or operational environment.

Introduction

In our own Soldiers and in the enemy combatants, control of stress is often the decisive difference between victory and defeat across the operational continuum. Battles and wars are won more by controlling the will to fight than by killing all of the enemy combatants. Uncontrolled combat stress causes erratic or harmful behaviors, impair mission performance, and may result in disaster and defeat of COSC preventive measures. The COSC preventive measures are aimed at minimizing maladaptive stress reactions while promoting adaptive stress reactions, such as loyalty, selflessness, and acts of bravery.

This manual provides doctrinal guidance for controlling excessive stress in combat and other operational environments. It identifies command and leadership responsibilities for COSC. It identifies COSC consultation, training, and education assistance available for units. This manual provides definitive guidance to BH personnel and CSC units for their COSC mission and for management of COSR and other behavioral disordered patients (BDPs). It identifies the requirements for COSC consultation, planning, coordination, rehearsal, and implementation of the COSC plan contained in the FHP annex of the operation order (OPORD).

Many stressors in a combat situation are due to deliberate enemy actions aimed at killing, wounding, or demoralizing our Soldiers and our allies. Other stressors are due to the operational environment. Some of these stressors can be avoided or counteracted by wise command actions. Still other stressors are due to our own calculated or miscalculated choice, accepted in order to exert greater stress on the enemy. Sound leadership works to keep stressors within tolerable limits and prepares the troops mentally and physically to endure them. Some of the most potent stressors can be due to personal organizational problems in the unit or on the home front. These, too, must be identified and, when possible, corrected or controlled. Unit needs assessments (UNAs) can help BH providers identify specific stressor in a unit and develop interventions to help unit personnel cope.

This manual identifies MH sections and medical units, CSC, that provides COSC support to units. It identifies the requirements for COSC assistance, as a unit transitions through the different phases of a deployment. Chaplains, commander, leader and COSC Soldier mentors provide assistance with after-action review (AAR) and traumatic events management (TEM).

Chapter 1

Combat and Operational Stress Control

SECTION I — INTERVENTION AND CONTROL FOR THE COMBAT AND OPERATIONAL STRESS THREAT

STRESS CONTROL

1-1. Combat stress includes all the physiological and emotional stresses encountered as a direct result of the dangers and mission demands of combat, see AR 40-216. Combat and operational stress control in the Army may be defined as programs developed and actions taken by military leadership to prevent, identify, and manage adverse COSRs in units. This program optimizes mission performance; conserves the fighting strength; and prevents or minimizes adverse effects of COSR on Soldiers and their physical, psychological, intellectual, and social health. Its goal is to return Soldiers to duty expeditiously. According to DODD 6490.2, COSC activities include routine screening of individuals when recruited; continued surveillance throughout military service, especially before, during, and after deployment; continual assessment and consultation with medical and other personnel from garrison to the battlefield.

1-2. Combat and operational stress control is the commander's responsibility at all levels. The commander is assisted with his responsibility for COSC by his staff, unit leaders, unit chaplain (Appendix A), and organic medical personnel. The commander may also receive assistance from organic COSC personnel at brigade and above, and from corps and above medical company/detachment CSC BH personnel. The key concern to combat commanders is to maximize the return-to-duty (RTD) rate of Soldiers who are temporarily impaired or incapacitated with stress-related conditions or diagnosed behavioral disorders.

1-3. The purpose of COSC is to promote Soldier and unit readiness by—

- Enhancing adaptive stress reactions.
- Preventing maladaptive stress reactions.
- Assisting Soldiers with controlling COSRs.
- Assisting Soldiers with behavioral disorders.

Note. The word control is used with combat and operational stress, rather than the word management, to emphasize the active steps that leaders, supporting BH personnel, and individual Soldiers must take to keep stress within an acceptable range.

COMBAT AND OPERATIONAL STRESS THREAT

1-4. Many stressors in a combat situation are due to deliberate enemy actions aimed at killing, wounding, or demoralizing our Soldiers and our allies. Other stressors are due to the natural environment, such as intense heat or cold, humidity, or poor air quality. Still others are due to leaders' own calculated or miscalculated choices (for example, decisions about unit strength, maneuver, the time of the attack, and plans for medical and logistical support). Sound leadership works to keep operational stressors within tolerable limits and prepares troops mentally and physically to endure them. In some cases however, excessive stress can affect both leaders' and Soldiers' decision-making and judgment, resulting in missed opportunities, or worse, in high casualties and/or failure to complete the mission. Finally, some of the most potent stressors are interpersonal in nature and can be due to conflict in the unit or on the home front. Extreme reactions to such stressors may involve harm to self (as in the hypothetical case of a Soldier that becomes suicidal on discovering that his wife wants a divorce) or to others (as in the case of a Soldier that impulsively fires a weapon at his noncommissioned officer (NCO) out of rage over perceived unfairness). These stressors must be identified and when possible, corrected or controlled. For more information on the control of combat/operational stressors and for details about specific leader actions to control stress, see FM 22-51. Also see FM 4-02 and FM 4-02.17 for additional information on the general and medical threat.

EFFECT OF STRESS

1-5. Focused stress is vital to survival and mission accomplishment. However, stress that is too intense, or prolonged, results in COSR that impairs their ability to function effectively. Some stressors contribute to misconduct that requires disciplinary action and may take the Soldier from duty for legal action and incarceration. In a broader context stress may cause battle and nonbattle injuries through inattention, clumsiness, and reckless behavior. These resultant injuries can include equipment losses and friendly fire incidents. Stress may increase disease rates by disrupting hygiene and protective measures, and impairing the body's immune defenses. Stress may progress to BH disorders, or suicidal behaviors and/or homicidal behaviors. Excessive stress in combat contributes to lapses in operational and tactical judgment and to missed opportunities that could increase the numbers of Soldiers injured over time.

MODEL FOR COMBAT AND OPERATIONAL STRESS CONTROL INTERVENTIONS

1-6. Combat and operational stress control assessments are performed during all phases of combat operations, stability and reconstruction operations, and support operations so the term COSC may refer to any of these different types of Army operations. The COSC assessments are performed at unit and individual level. They consider the range of variables according to a model, which recognizes that biological, psychological, and social factors influence each other. The COSC assessment reviews the interaction systematically to a depth appropriate to the need. The assessment identifies which variables can be modified to improve coping or outcome. Based on these assessments COSC personnel recommend courses of action (COA) to the commander. They identify and initiate COSC interventions to improve unit effectiveness and Soldier efficiency and well-being.

1-7. Provided in Figure 1-1 is a conceptual model of stress, its mitigating and aggravating factors and potential outcomes on Soldiers and families. This model can be helpful when designing COSC interventions to improve short-term and long-term outcomes.

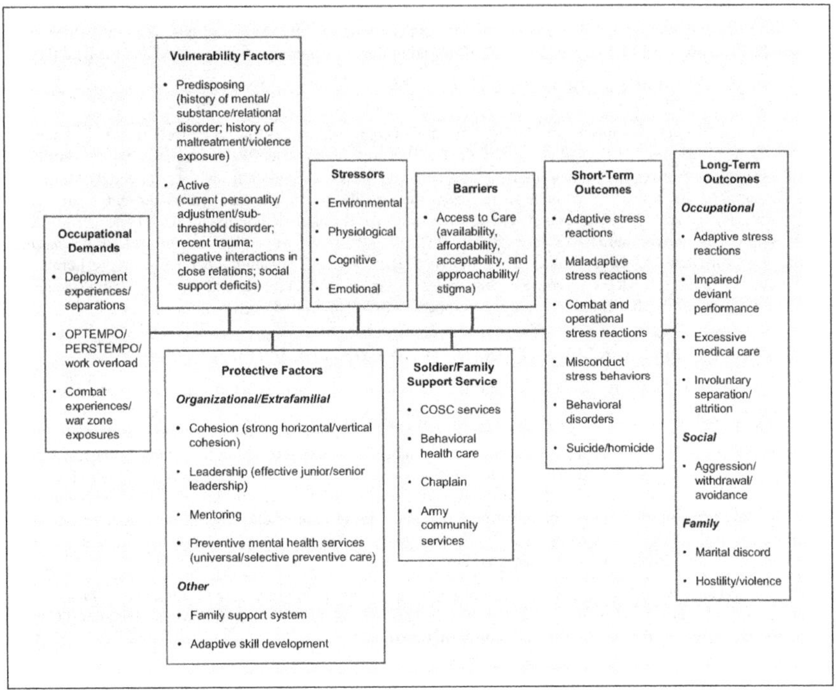

Figure 1-1. Model of stress and its potential Soldier and family outcomes

MENTAL AND PHYSICAL STRESSORS

1-8. A rough distinction between a mental and physical stressor can be made—

- A mental stressor is one in which the brain receives information about a given threat or demand, but this information results in only indirect physical impact on the body. Instead, its primary effect is to place demands on and evoke reactions from the perceptual, cognitive and/or emotional systems of the brain (such as information overload, perceived lack of control, or grief-producing losses).
- A physical stressor has a direct, potentially harmful effect on the body. These stressors may be external environmental conditions (such as temperature) or the internal physiologic demands required by or placed upon the human body (such as the need for hydration, or an immune response to a viral infection).

TYPES OF MENTAL AND PHYSICAL STRESSORS

1-9. Figure 1-2 provides examples for the two types of physical stressors (environmental and physiological) and the two types of mental stressors (cognitive and emotional). Also, as recognized in the COSC intervention model (Figure 1-1) physical stressors cause mental stressors when they result in discomfort, distraction, and threat of harm, as well as when they directly impair brain functions. Mental stressors can lead to adaptive or maladaptive stress behaviors that decrease or increase the exposure to physical stressors.

PHYSICAL STRESSORS	MENTAL STRESSORS
ENVIRONMENTAL	**COGNITIVE**
Heat, cold, wetness, dust Vibration, noise, blast Noxious odors (fumes, poisons, chemicals) Directed-energy weapons/devices Ionizing radiation Infectious agents Physical work Poor visibility (bright lights, darkness, haze) Difficult or arduous terrain High altitude	Information (too much or too little) Sensory overload or deprivation Ambiguity, uncertainty, unpredictability Time pressure or waiting Difficult decision (rules of engagement) Organizational dynamics and changes Hard choices versus no choice Recognition of impaired functioning Working beyond skill level Previous failures
PHYSIOLOGICAL	**EMOTIONAL**
Sleep deprivation Dehydration Malnutrition Poor hygiene Muscular and aerobic fatigue Overuse or underuse of muscles Impaired immune system Illness or injury Sexual frustration Substance use (smoking, caffeine, alcohol) Obesity Poor physical condition	Being new in unit, isolated, lonely Fear and anxiety-producing threats (of death, injury, failure, or loss) Grief-producing losses (bereavement) Resentment, anger, and rage-producing frustration and guilt Inactivity producing boredom Conflicting/divided motives and loyalties Spiritual confrontation or temptation causing loss of faith Interpersonal conflict (unit, buddy) Home-front worries, homesickness Loss of privacy Victimization/harassment Exposure to combat/dead bodies Having to kill

Figure 1-2. Examples of combat and operational stressors

STRESS BEHAVIORS IN COMBAT AND OTHER OPERATIONS

1-10. Combat and operational stress behavior is the generic term that is used for the full spectrum of combat and operational stress behaviors. It covers the range of reactions, from adaptive to maladaptive. Figure 1-3 provides a listing of typical adaptive and maladaptive stress reactions.

ADAPTIVE STRESS REACTIONS

1-11. Stressors, when combines with effective leadership and good peer relationships may lead to adaptive stress reactions which enhance individual and unit performance. Examples of adaptive stress reactions include—

- The strong personal trust, loyalty, and cohesiveness (called horizontal bonding), which develops among peers in a small military unit.
- Personal trust, loyalty, and cohesiveness that develops between leaders and subordinates (called vertical bonding).
- Esprit de corps is also defined as a feeling of identification and membership in the larger, enduring unit with its history and ideas. This may include the unit (such as battalion, brigade

combat team [BCT], regiment, or division) the branch (such as infantry, artillery, or military police [MP]) and beyond the branch to the US Army level.

- Unit cohesion is the binding force that keeps Soldiers together and performing the mission in spite of danger and death. Cohesion is a result of Soldiers knowing and trusting their peers and leaders and understanding their dependency on one another. It is achieved through personal bonding and a strong sense of responsibility toward the unit and its members. The ultimate adaptive stress reactions are acts of extreme courage and almost unbelievable strength. They may even involve deliberate self-sacrifice.

MALADAPTIVE STRESS REACTIONS

1-12. Combat and operational stress reaction and misconduct stress behaviors comprise the maladaptive stress reactions.

COMBAT AND OPERATIONAL STRESS REACTION

1-13. The Army uses the DOD-approved term/acronym COSR in official medical reports. This term can be applied to any stress reaction in the military unit environment. Many reactions look like symptoms of mental illness (such as panic, extreme anxiety, depression, hallucinations), but they are only transient reactions to the traumatic stress of combat and the cumulative stresses of military operations. Some individuals may have behavioral disorders that existed prior to deployment or disorders that were first present during deployment, and need BH intervention beyond the interventions for COSR.

MISCONDUCT STRESS BEHAVIORS

1-14. Examples of misconduct stress behaviors are listed in Figure 1-3. These range from minor breaches of unit orders or regulations to serious violations of the Uniform Code of Military Justice (UCMJ) and the Law of Land Warfare. Misconduct stress behaviors are most likely to occur in poorly trained undisciplined Soldiers, however good and heroic, under extreme combat stress may also engage in misconduct. Generally, misconduct stress behaviors—

- Range from minor breaches of unit orders or regulations to serious violations of the UCMJ and the Law of Land Warfare.
- May also become a major problem for highly cohesive and proud units. Such units may come to consider themselves entitled to special privileges and, as a result, some members may relieve tension unlawfully when they stand-down from their military operations. They may lapse into illegal revenge when a unit member is lost in combat. Stress control measures and sound leadership can prevent such misconduct stress behaviors, but once serious misconduct has occurred, Soldiers must be punished to prevent further erosion of discipline. Combat stress, even with heroic combat performance, cannot justify criminal misconduct.

OVERLAPPING STRESS BEHAVIORS

1-15. The distinctions among adaptive stress reactions, misconduct stress behaviors, and COSRs are not always clear. Indeed, the three categories of combat and operational stress behaviors may overlap. Soldiers with COSR may show misconduct stress behaviors and vice versa. Soldiers with adaptive stress reactions may also suffer from COSR. Finally, excellent combat Soldiers that have exhibited bravery and acts of heroism may also commit misconduct stress behaviors.

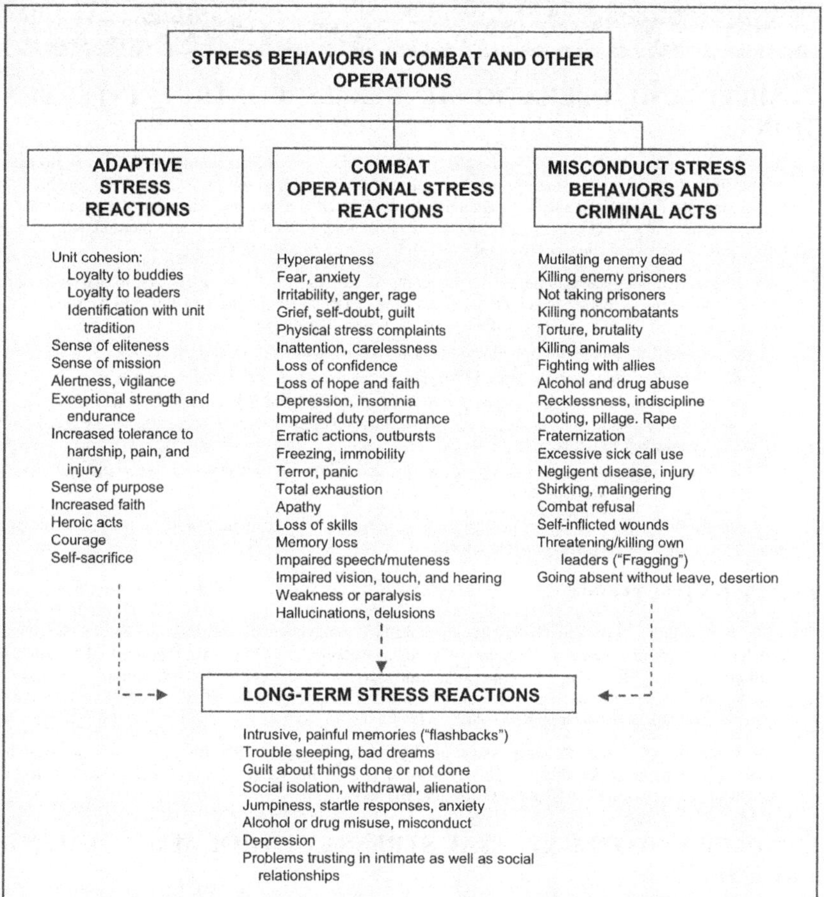

Figure 1-3. Stress behaviors in combat and other operations

1-16. It is common for stress reactions to persist or arise long after exposure to distressing events. When there is impairment in social and/or occupational functioning, a clinical assessment is warranted. Combat and operational stress control is important to sustain Army strength over the long-term and reduce the cost to society, the DOD, the Soldiers, and families.

SECTION II — PRINCIPLES AND FUNCTIONAL AREAS FOR COMBAT AND OPERATIONAL STRESS CONTROL

COMBAT AND OPERATIONAL STRESS CONTROL INTERVENTIONS

1-17. Soldier and unit readiness is best achieved through an active, prevention-focused orientation that is embodied in the principles. These principles apply to all COSC interventions or activities throughout the theater and are followed by medical personnel in all BH/COSC elements. Their application may differ based on a particular level of care and other factors pertaining to the mission, enemy, terrain and weather, troops and support available, time available, and civil considerations (METT-TC).

1-18. Preventive interventions seek to reduce the occurrence or severity of COSR and behavioral disorders, thereby sustaining Soldier and unit readiness. These interventions are tailored to the needs of the population. There are four categories of preventive interventions that include—

- Universal. Interventions targeted to the general population or an assigned AO.
- Selective. Interventions targeted to a unit or Soldier whose risk is higher than average.
- Indicated. Interventions targeted to Soldiers with COSR or indications of a potential behavioral disorder, and to units that show signs their mission effectiveness is being affected by combat and operational stressors.
- Treatment. Interventions targeted to treat and follow-up Soldiers with behavioral disorders to prevent their loss from duty.

1-19. Identify life- or function-threatening medical, surgical, or psychiatric condition as soon as possible and provide those patients emergency treatment.

COMMUNICATIONS

1-20. To maximize prevention, COSC personnel must maintain a high degree of involvement with Soldiers and leaders of supported units. Regular visits to the battalion/squadron level are essential to maintain appropriate accessibility. The COSC personnel must always be attuned to Soldiers around them and not rely on Soldiers to come to them. This means that when time permits, COSC personnel need to start conversations with as many supported Soldiers as possible.

1-21. Communication is the delivery system of COSC services and therefore is essential. Use all available means to communicate/coordinate with supported units, subordinate teams, and higher headquarters to ensure delivery of effective and timely COSC and BH services.

COMBAT AND OPERATIONAL STRESS CONTROL MANAGEMENT PRINCIPLES

1-22. The COSC management principles of brevity, immediacy, contact, expectancy, proximity, and simplicity (BICEPS). These principles apply to all COSC interventions or activities throughout the theater, and are followed by COSC personnel in all BH/COSC elements. These principles may be applied differently based on a particular level of care and other factors pertaining to METT-TC.

> ***Note.*** The BICEPS principles are discussed in-depth in FM 6-22.5. This is a multiservice publication with the US Marine Corps (USMC). Due to differences in organizational structure the USMC defines the "C" in BICEPS to mean *centrality*, while the US Army defines it to mean *contact*, which will be used for the remainder of this manual.

BREVITY, IMMEDIACY, CONTACT, EXPECTANCY, PROXIMITY, AND SIMPLICITY

1-23. Using BICEPS is extremely important in the management of Soldiers with COSR and/or behavioral disorders.

Brevity

1-24. Initial rest and replenishment at COSC facilities located close to the Soldier's unit should last no more than 1 to 3 days (USMC and Navy is 3 to 4 days). Those requiring further treatment are moved to the next level of care. Since many require no further treatment, military commanders expect their Soldiers to RTD rapidly.

Immediacy

1-25. It is essential that COSC measures be initiated as soon as possible when operations permit. Intervention is provided as soon as symptoms appear.

Contact

1-26. The Soldier must be encouraged to continue to think of himself as a warfighter, rather than a *patient* or a *sick person*. The chain of command remains directly involved in the Soldier's recovery and RTD. The COSC team coordinates with the unit's leaders to learn whether the overstressed individual was a good performer prior to the COSR. Whenever possible, representatives of the unit or messages from the unit tell the Soldier that he is needed and wanted back. The COSC team coordinates with the unit leaders, through unit medical personnel or chaplains, any special advice on how to assure quick reintegration when the Soldier returns to his unit.

Expectancy

1-27. The individual is explicitly told that he is reacting normally to extreme stress and is expected to recover and return to full duty in a few hours or days. A military leader is extremely effective in this area of treatment. Of all the things said to a Soldier suffering from COSR the words of his small-unit leader have the greatest impact due to the positive bonding process that occurs. A simple statement from the small-unit leader to the Soldier that he is reacting normally to COSR and is expected back soon have positive impact. Small-unit leaders should tell Soldiers that their comrades need and expect them to return. When they do return, the unit treats them as every other Soldier and expects them to perform well.

Proximity

1-28. Soldiers requiring observation or care beyond the unit level are evacuated to facilities in close proximity to, but separate from the medical or surgical patients at the battalion aid station (BAS) or medical company nearest the Soldiers' unit. It is best to send Soldiers who cannot continue their mission and require more extensive intervention to a facility other than a hospital, unless no other alternative is possible. Combat and operational stress reactions are often more effectively managed in areas close to the Soldier's parent unit. On the noncontiguous battlefield characterized by rapid, frequent maneuver and continuous operations, COSC personnel must be innovative and flexible in designing interventions which maximize and maintain the Soldier's connection to his parent unit. See FM 6-22.5 for additional definition of proximity.

Simplicity

1-29. Indicates the need to use brief and straightforward methods to restore physical well-being and self-confidence.

1-30. The actions used for COSR control (commonly referred to as the 5 R's) involve the following actions:

- **R**eassure of normality.
- **R**est (respite from combat or break from the work).

- **R**eplenish bodily needs (such as thermal comfort, water, food, hygiene, and sleep).
- **R**estore confidence with purposeful activities and contact with his unit.
- **R**eturn to duty and reunite Soldier with his unit.

Note. Historically the Army had used the terms proximity, immediacy, expectancy and simplicity (PIES) but began using BICEPS when it became the approved joint terminology.

NONPATIENT STATUS

1-31. To prevent Soldiers with COSR from adopting the *patient* role, these guidelines should be followed:

- Keep the Soldier in uniform and hold him responsible for maintaining Soldier standards.
- Keep the Soldier separate from seriously ill or injured patients.
- Avoid giving him medications unless essential to manage sleep.
- Do not evacuate or hospitalize the Soldier unless absolutely necessary.
- Do not diagnose the Soldier prematurely.
- Transport the Soldier via general-purpose vehicles, not ambulances.

COMBAT AND OPERATIONAL STRESS CONTROL FUNCTIONAL AREAS

1-32. Combat and operational stress control interventions and activities are organized into nine functional areas. These functional areas cover the full spectrum of BH care from preventive through clinical intervention. They are defined below and are discussed further in subsequent chapters of this manual.

UNIT NEEDS ASSESSMENT

1-33. Unit needs assessment is the systematic and frequent assessment of supported units to determine the priority and types of BH interventions required. See Chapter 4.

CONSULTATION AND EDUCATION

1-34. Consultation involves the liaison with and preventive advice to commanders, staff of supported units, and Soldiers. Education involves training in concepts and skills for increasing Soldier resilience to stress. See Chapter 5.

TRAUMATIC EVENTS MANAGEMENT

1-35. Traumatic events management blends other COSC functional areas to create a flexible set of interventions specifically focused on stress management for units and Soldiers following potentially traumatizing events (PTE). Like other functional areas, COSC providers must tailor TEM to the needs of the unit and the Soldier. See Chapter 6.

RECONSTITUTION SUPPORT

1-36. Reconstitution is extraordinary action that commanders plan and implement to restore units to a desired level of combat effectiveness commensurate with mission requirements and available resources. It transcends normal day-to-day force sustainment actions. However, it uses existing systems and units to do so. No resources exist solely to perform reconstitution. In COSC reconstitution support, COSC personnel are responsible for providing unit Soldier restoration and conducting the COSC functions. This support is provided to units following traumatic events and during reconstitution, redeployment, and transition among levels of operational tempo (OPTEMPO). See Chapter 7.

Combat and Operational Stress Control Triage

1-37. Combat and operational stress control triage is the process of sorting Soldiers with COSR and/or BH disorders based upon where they can best be managed. See Chapter 8.

Combat and Operational Stress Control Stabilization

1-38. This function provides initial management of Soldiers with severe COSR or behavioral disorders. Their safety is ensured and they are evaluated for RTD potential or prepared for further treatment or evacuation. See Chapter 9.

Soldier Restoration

1-39. Soldier restoration involves the one- to three-day management of Soldiers with COSR or behavioral disorders normally near a medical treatment facility (MTF) in close proximity to his unit. This approach uses the 5 R's discussed above. See Chapter 10.

Behavioral Health Treatment

1-40. Patients with identified behavioral disorders receive ongoing evaluation, treatment and follow-up to sustain them. This functional area implies a therapist-patient relationship, clinical documentation, and adherence to clinical standards of care. See Chapter 11.

Soldier Reconditioning

1-41. Reconditioning is an intensive program of work therapy, military activities, physical training and psychotherapy. Reconditioning programs are conducted up to seven days (or more) in the corps area. Additional reconditioning may be provided in the theater outside the combat zone (CZ). See Chapter 12.

Priority of Functional Areas

1-42. The medical company/detachment, CSC commanders, MH section leaders, senior medical commanders, and command surgeons must set priorities, coordinate their actions, and allocate resources to accomplish missions based upon the total situation/METT-TC. Needs assessments and available resources guide COSC interventions. At the unit level, the unit needs assessment helps determine which functional areas (such as consultation and education; and critical event and transition management) take priority to achieve success with the COSC mission. At the Soldier level, COSC triage helps determine which functional areas (such as consultation and education, stabilization, Soldier restoration, behavioral health treatment, and reconditioning) take priority to achieve COSC mission goals.

Chapter 2

Behavioral Health and Combat and Operational Stress Control Elements in the Theater

SECTION I — MENTAL HEALTH SECTIONS

ASSIGNMENTS

2-1. Mental health sections are located in medical companies assigned to brigades, divisions, corps, and theater-level medical units.

ARMY OF EXCELLENCE

2-2. In the Army of Excellence (AOE) divisions, the division MH section (DMHS) is organic to the main support medical company (MSMC). In the armored cavalry regiments (ACRs) and separate brigades, a MH section is assigned to the forward support medical company (FSMC) of the forward support battalion (FSB) or a medical troop. In the corps and echelons above corps (EAC), the MH sections are located in the headquarters and headquarters detachment (HHD), area support medical battalion (ASMB) and in the area support medical companies (ASMCs). In those AOE divisions that are undergoing transformation, the MH sections are assigned to each of the brigade support medical companies (BSMCs) of the brigade support battalion (BSB) and in the ASMCs from the corps ASMB.

MODULAR FORCE

2-3. In the modular BCTs, Stryker BCTs (SBCTs) and the heavy BCTs (HBCTs) MH sections are assigned to the BSMC of the BSB. At echelons above brigade (EAB) and at the division, corps, or theater levels, the MH sections are assigned to the multifunctional medical battalion (MMB).

> ***Note.*** Under the modular design, the MMB will replace the ASMB, the medical logistics battalion, and the medical evacuation battalion.

ORGANIZATIONS

2-4. The AOE divisions have only one MH section that is assigned to the MSMC of the main support battalion (MSB), which is located in the division support area (DSA). See Table 2-1 for a listing of personnel assigned to the division MH section. These personnel are deployed to the supported brigade as two-man teams comprised of one officer and one MH specialist.

Table 2-1. Division mental health staff

Division Psychiatrist (MAJ, Area of Concentration [AOC] 60W00, Medical Corps [MC])
Social Work Officer (CPT, AOC 73A67, Medical Service [MS] Corps)
Clinical Psychologist (CPT, AOC 73B67, MS)
Mental Health NCO (E7, E6, and E5, Military Occupational Specialty [MOS] 68X40/30/20)
Mental Health Specialist (Four)—(Two E4 and Two E3, MOS 68X10)

FUNCTIONS AND RESPONSIBILITIES OF ALL MENTAL HEALTH SECTIONS

2-5. All MH sections regardless of their organizational assignment are tasked with providing COSC for their supported units. In all of these units, COSC is accomplished through vigorous prevention, consultation, training, educational, and Soldier restoration programs. These programs are designed to provide BH expertise to unit leaders and Soldiers where they serve and sustain their mission focus and effectiveness under heavy and prolonged stress. The MH sections identify Soldiers with COSRs who need to be provided rest/Soldier restoration within or near their unit area for rapid RTD. These programs are designed to maximize the RTD rate of Soldiers who are either temporary impaired, have a diagnosed behavioral disorder, or have stress-related conditions. Also the prevention of posttraumatic stress disorders (PTSDs) is an important objective for brigades and EAB. The behavioral health officer (either a clinical psychologist or social work officer) and MH specialist are especially concerned with assisting and training of—

- Small-unit leaders.
- Unit ministry teams (UMTs) and staff chaplains.
- Battalion medical platoons.
- Patient-holding squad and treatment squad personnel of the medical company.

RESPONSIBILITIES

2-6. The MH section has a primary responsibility for assisting commanders with COSC by implementing the brigade combat mental fitness program. Also the MH section serves as a consultant to the commander, staff, and others involved with providing prevention and intervention services to unit Soldiers and their families. The MH section has staff responsibilities for assisting the brigade surgeon with establishing brigade policy and guidance for the prevention, diagnosis, treatment, management, and RTD of stress-related casualties. This is accomplished under the guidance and in close coordination with maneuver battalions and FSMC/BSMC physicians.

FUNCTIONS

2-7. The COSC functional areas were identified in Chapter 1. The functions of all MH sections are to support the nine functional areas except Soldier reconditioning.

UTILIZATION IN GARRISON

2-8. In garrison, BH personnel assigned to the division, brigade, or corps/EAC units continue to perform the same staff and outreach functions with supported units as they do in a field environment. An increase in the BH treatment functions may be possible as a result of consolidating BH care providers. The BH providers make available their consultation skills and clinical expertise to the Soldiers of supported units and their family readiness groups. Clinical care of family members and of Soldiers that require longer-term care beyond crisis intervention, brief treatment, and medication follow-up is the responsibility of the medical department activity (MEDDAC)/medical center (MEDCEN). The MH section personnel should focus their clinical work primarily on Soldiers with problems amenable to brief treatment.

2-9. Clinical services may be provided as part of a consolidated BH activity (as is usually established by a DMHS) or by augmenting an existing MEDDAC/MEDCEN BH staff.

2-10. The MH sections should strive to reduce referrals for BH treatment by working closely with unit leaders and chaplains to control organizational stress and rapidly identify and intervene with those Soldiers having BH disorders.

2-11. When the medical company or battalion deploys on training exercises, assigned BH personnel deploy with them to provide COSC training and support. In addition, they train to improve their own technical and tactical skills.

Note. In accordance with AR 40-216, clinical responsibilities in garrison must not interfere with participation in field and deployment exercises and maintenance of combat readiness.

MEDICAL DETACHMENT, COMBAT STRESS CONTROL

MISSION

2-12. The mission of the medical detachment, CSC (TOE 08463A000) is to provide COSC interventions and stress prevention activities to supported units in its AO. The staff augments division and brigade MH sections; provides direct support (DS) to combat brigades without organic BH officers; and provides area support in its AO. The medical detachment, CSC reconstitutes other brigade and division COSC assets. The medical detachment, CSC provides COSC interventions and activities to indigenous populations as directed in stability and reconstruction operations, humanitarian assistance, disaster relief, peace support operations, and detention facility operations. The medical detachment, CSC provides COSC interventions and activities to units in support of their readiness preparation and throughout their deployment cycle.

ASSIGNMENT

2-13. The Medical Reengineering Initiative (MRI) medical detachment, CSC (TOE 08463A000) is assigned to a corps medical brigade (MEDBDE) (TOE 08422A100), or other medical command and control (C2) element. Its teams may be further attached to a medical company, CSC (TOE 08467L000), an ASMB (TOE 08456A000), a BSMC/MSMC, or a combat support hospital (CSH). See Appendix B for the proposed future medical detachment, COSC.

EMPLOYMENT AND CAPABILITIES

2-14. The detachment headquarters is usually located in the DSA. Its teams disperse throughout the division's AO, and may extend support to corps if no CSC medical company is available. At TOE Level 1, the MRI medical detachment, CSC provides—

- A preventive section capable of dividing into four preventive teams. Each team provides mobile COSC interventions and activities.
- A fitness section capable of dividing into two fitness teams. Each fitness team provides mobile COSC interventions and activities. Each team is equipped to hold 40 Soldiers at the same time. With additional logistical support, each team can accommodate additional Soldiers under surge conditions.

STAFF RESPONSIBILITIES

2-15. The medical detachment, CSC assists the C2 headquarters (of the unit to which it is assigned or attached) regarding planning and coordination of COSC support, stress threat, mental and physical stressors, stress behaviors, principles of COSC, and implementation of COSC functional areas.

DEPENDENCY

2-16. This unit is dependent on—

- Appropriate elements of the corps for FHP, religious support, legal, finance, field feeding, personnel and administrative services support; laundry and clothing exchange; mortuary affairs support; and security of enemy prisoners of war (EPW), detainee, and US prisoner patients.
- The medical headquarters to which it is assigned/attached for FHP; medical administration, logistics, including MEDLOG; medical regulating of patients; evacuation; coordination for RTD; and unit level equipment and communications-electronics (CE) maintenance.

Mobility and Security

2-17. This unit is 100-percent mobile. It requires 100 percent of its organic personnel and equipment be transported in a single lift, using its organic vehicles. Upon relocation, Soldiers being held will require additional transportation. This unit is responsible for perimeter defense of its immediate operational area. However, it is dependent on appropriate elements of the corps for additional security, to include security of convoy operations. Personnel of the detachment are provided weapons for their personal defense and for the defense of their patients and/or held Soldiers.

ORGANIZATION

2-18. This 43-person MRI medical detachment (Figure 2-1) is organized into a headquarters section; a preventive section composed of four CSC preventive (CSCP) teams; and a fitness section consisting of two CSC fitness (CSCF) teams.

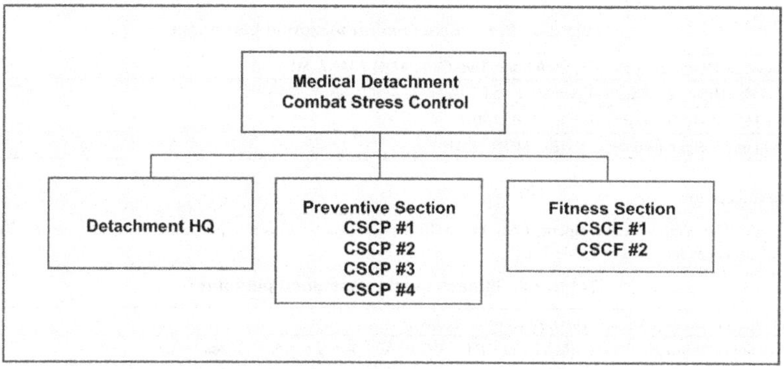

Figure 2-1. Medical Reengineering Initiative medical detachment, combat stress control (TOE 08463A000)

Detachment Headquarters

2-19. The medical detachment, CSC headquarters provides C2 for the detachment. The headquarters element is responsible for planning, coordinating, and implementing COSC support for supported units. Personnel of the headquarters element provide maintenance, supply and service, and personnel administrative support. See Table 2-2 for a list of personnel assigned to the detachment headquarters.

Table 2-2. Detachment headquarters assigned personnel

Detachment Commander (LTC, AOC 60W00, MC)
Field Medical Assistant (1LT, AOC 70B67, MS)
Detachment Sergeant (E7, MOS 68X40)
Supply Sergeant (E5, MOS 92Y20)
Human Resource Specialist (E4, MOS 42A10)
Wheeled Vehicle Mechanic (E4, MOS 63B10)
Cook (E3, MOS 92G10)
Note. Detachment officers and NCOs from the preventive section and the fitness section may be assigned additional duties, which enhance the overall effectiveness of the headquarters section.

Preventive Section

2-20. The medical detachment, CSC has a 16-person preventive section. See Table 2-3 for listing of personnel assigned to the preventive section. The section can be divided into four separate CSCP teams. Three CSCP teams are normally allocated to supported maneuver brigades (one team per brigade). The fourth CSCP team may provide direct support to a SBCT, an aviation brigade, other brigade-sized units, or to corps units operating in a division's AO. Preventive section personnel may be task-organized with personnel of the fitness section into teams for specific missions. The section (and team) leader position may be held by any of the officers assigned to the section. The section's COSC interventions and activities are—

- Unit needs assessment; consultation and education; TEM; COSC triage; stabilization (emergency); and BH treatment.
- Assisting with Soldier restoration and reconditioning at the CSC detachment program. Overseeing a one- to three-day COSC Soldier restoration program in a brigade, division, or ASMC holding section or in another area suitable for Soldiers experiencing COSR and/or other stress-related disorders.

Table 2-3. Preventive section assigned personnel

Social Work Officer (Four)—(Two MAJ and Two CPT, AOC 73A67, MS)
Clinical Psychologist (Four)—(Two MAJ and Two CPT, AOC 73B67, MS)
Mental Health NCO (Four)—(E5, MOS 68X20)
Mental Health Specialist (Four)—(E4, MOS 68X10)

Fitness Section

2-21. The medical detachment, CSC has a 20-person fitness section. Personnel assigned to the fitness section are identified in Table 2-4.

Table 2-4. Fitness section assigned personnel

Psychiatrist (Two)—(LTC and MAJ, AOC 60W00, MC)
Occupational Therapist (Two)—(MAJ and CPT, AOC 65A00, Army Medical Specialty Corps [SP])
Psychiatric/BH Nurse (Two)—(MAJ and CPT, AOC 66C00, Army Nurse Corps [AN])
Occupational Therapy NCO (Four)—(Two E6 and Two E5, MOS 68W30/20N3)
Mental Health NCO (Four)—(Two E6 and Two E5, MOS 68X30/20)
Mental Health Specialist (Six)—(Four E4 and Two E3, MOS 68X10)

2-22. The fitness section is task-organized to provide COSC interventions and activities. Fitness section personnel may be task-organized with personnel of the preventive section into teams for specific missions. This section can divide into two CSCF teams. Each fitness team can deploy a four-person mobile team using the team's high-mobility multipurpose wheeled vehicle (HMMWV). One CSCF team usually collocates with a supported divisional medical company to provide mobile COSC support within a DSA and conduct Soldier restoration programs, as required. The second CSCF team augments area support to corps units in the division AO and in forward areas of the corps. The CSCF team provides staff and equipment for operating a Soldier restoration or reconditioning center. This section's COSC interventions and activities are—

- Unit needs assessment; consultation and education; TEM; COSC triage; stabilization; and BH treatment.
- Conducting Soldier restoration and reconditioning programs.
- Staffing and operating a psychiatric ward for supported CSH when this capability is required (refer to FM 8-10-14).

MEDICAL COMPANY, COMBAT STRESS CONTROL

2-23. The MRI medical company, CSC is employed in the corps and EAC. The basis of allocation is one medical company, CSC per corps or theater. The medical company, CSC is task-organized, METT-TC-dependent for stability and reconstructions operations. Medical company, CSC (TOE 08467A000) replaces medical company, CSC (TOE 08467L000). A medical company, CSC provides comprehensive preventive and treatment services to a corps and EAC during war. It provides this support to all services on an area support basis. The medical company, CSC provides DS to combat and combat support (CS) brigades, as needed. It reinforces or reconstitutes other CSC assets in the corps or divisions if required. The medical company, CSC provides COSC/BH services to indigenous populations as directed in stability and reconstruction operations, humanitarian assistance, disaster relief, and peace support operations. The comprehensive support provided by the CSC medical company entails all of the nine COSC functional mission areas. The COSC functional mission areas are identified in Chapter 1 and discussed in subsequent chapters of this publication.

CAPABILITIES

2-24. At TOE Level 1, the CSC medical company provides—

- Advice, planning, and coordination for COSC to commanders.
- Reconstitution (COSC) support for units up to division size.
- Preventive and fitness teams (four to ten personnel) for consultation, treatment services, and reconstitution support for up to battalion-sized organizations.
- Soldier restoration or reconditioning programs for up to 50 Soldiers per fitness team on an area basis.
- Deployment of COSC elements to forward areas for support of contingency operations.

ASSIGNMENT

2-25. The medical company, CSC (TOE 08467A000) is assigned to a Medical Command (US Army) (MEDCOM). Elements of this TOE may be further attached to a corps MEDBDE or to either a MMB or an ASMB.

ORGANIZATION

2-26. The medical company, CSC is organized into a headquarters section, a preventive section, and a fitness section. The company is dependent on appropriate elements of the MEDCOM or MEDBDE for administrative and MEDLOG support, medical regulating, COSR casualty delivery, and medical evacuation. The company is dependent on appropriate elements of the corps or EAC for finance, legal, personnel and administrative services, food service, supply and field services, supplemental transportation, and local security support services. When CSC medical company elements or teams are deployed to division areas, they are dependent on the division medical companies (such as MSMC or FSMC/BSMC) for patient accounting, transportation, food service, and field service support.

Employment

2-27. The medical company, CSC is employed in all intensities of conflict when a corps with two or more divisions is deployed. Task-organized COSC elements are deployed for division-sized combat operations, stability and reconstruction operations, and other contingency operations which are METT-TC-dependent. Together, the preventive and fitness sections provide all five BH disciplines. These resources are flexibly task-organized in a variety of combinations to meet the fluid COSC threat at different phases in the operations. Personnel may be quickly cross-attached from one section to another to accommodate the shifting workload and to provide reconstitution support packages. The preventive and fitness sections both organize into teams. The COSC preventive or fitness teams deployed forward of the corps boundaries in support of tactical operations come under the control of the FHP operations element in the supported units. These teams will also come under the operational control of the division or brigade CSC teams. One or more of the medical company, CSC's eight preventive teams may locate at the BSMC when deployed in DS of SBCT, HBCT, and separate brigades or ACRs.

Fitness Teams

2-28. One or more of the four (ten-person) fitness teams may reinforce ASMCs which are deployed to locations throughout the corps and EAC. These teams provide a basis for COSC prevention and intervention. The teams may conduct Soldier restoration programs at the ASMCs, as required. These teams may also be deployed forward to provide temporary augmentation/reinforcement, as required.

2-29. Based on workload, one or more of the four fitness teams, plus one or more preventive teams locate with a theater hospital where they conduct Level IV COSC reconditioning programs, as required. A hospital located in the corps rear or out of the CZ is the best location to conduct the theater COSC reconditioning program. When deployed with a hospital, these teams provide mobile consultation in the vicinity of the hospital. These teams are also prepared to restrict reconditioning programs and deploy forward in support of higher priority missions on very short notice. These teams can also augment hospital BH ward services by staffing a temporary BH ward.

2-30. The medical company, CSC is divisible into four functionally emulative increments for split-based operations and stability and reconstruction operations as assigned. Nonstandard task elements for specific missions can be organized using any combination of the preventive section and fitness section personnel to meet specific BH needs. For stability and reconstruction operations, the minimum is an officer/NCO team to supplement a brigade COSC team or a preventive module/team of two officers, one NCO, and one enlisted. These modules may be augmented with personnel from the fitness section to add additional specialty expertise.

Headquarters Section

2-31. The headquarters section provides C2 and unit-level administrative and maintenance support to its subordinate sections when they are collocated with the company. The headquarters section may also provide assistance to detached elements by making site visits if the elements are within a feasible distance for ground transportation. The CSC medical company elements normally deploy with limited maintenance capability. When these COSC elements deploy, they are dependent on the supported units for patient accounting, transportation, food service, and field services. The personnel assigned to the headquarters section are identified in Table 2-5. See Appendix C for information on the AOE L-Series TOE medical company and medical detachment, CSC.

Table 2-5. Medical company, combat stress control, headquarters section personnel

Commander (LTC, AOC 60W00, MC)
Chaplain (CPT, AOC 56A00, Chaplain [CH])
Field Medical Assistant (CPT, AOC 70B67, MS)
First Sergeant (E8, MOS 68W50)
Mental Health NCO (E7, MOS 68X40)
Chemical Operations Specialist (E5, MOS 54B20)
Patient Administrative NCO (E5, MOS 68G20)
Supply Sergeant (E5, MOS 92Y20)
Power Generator Equipment Repairer (E4, MOS 52D10)
Wheeled Vehicle Mechanic (E4, MOS 63B10)
Administrative Specialist (E4, MOS 71L10)
Patient Administrative Specialist (Two)—(E4 and E3, MOS 68G10)
Armorer (E4, MOS 92Y10)
Cook (E3, MOS 92G10)
Note. Personnel from the headquarters section are deployed with teams or task-organized COSC element, as required.

SECTION II — OPERATIONAL STRESS ASSESSMENT TEAM IN A THEATER

MISSION OF OPERATIONAL STRESS ASSESSMENT TEAMS

2-32. The operational stress assessment team (OSAT) is task-organized when needed from personnel in the research institutes of United States Army Medical Research and Materiel Command (USAMRMC). The mission of this team is to deploy with BH research expertise, equipment and supplies to provide scientifically valid needs assessment of multiple units in the theater and macro analysis of COSC stressors, stress, and Soldier needs. The source is predominantly the Walter Reed Army Institute of Research, augmented when needed with research psychologists and enlisted technicians from the US Army Aeromedical Research Laboratory, the US Army Medical Research Institute of Chemical Defense, and the US Army Research Institute of Environmental Medicine. The OSATs are USAMRMC teams that are assembled entirely from personnel in table of distribution and allowances (TDA) units and are dependent on TOE units when they deploy to a theater.

ASSIGNMENT AND CAPABILITIES

2-33. The OSAT normally is used for augmenting the senior medical command in the theater. The OSAT has the capabilities to—

- Plan, coordinate, and conduct questionnaire and interview surveys.
- Establish a database and analyze scientific-quality data from their own surveys and from standardized unit assessments conducted by unit MH section and CSC units for their local supported units.
- Provide rapid reports of findings to the command that include recommendations for policy and for COSC actions and interventions.
- Design and validate survey instruments, including ones to address new and emerging stressors and issues in the theater.
- Use USAMRMC capability to archive and analyze data for comparison with other conflicts.
- Educate and mentor COSC and other personnel in survey methodology and provide them, as well as the theater commander, the wider perspective to use in their local UNA.

- Interface with the theater surveillance capabilities of preventive medicine (PVNTMED) and occupational and environmental medicine to address the BH aspects of chemical, biological, radiological and nuclear (CBRN) exposures and threats.

DEPENDENCY

2-34. The OSAT depends on the medical headquarters or the units with which it is working for administrative support. The OSAT depends on the theater (line) command for implementation of the approved surveys.

MOBILITY AND SECURITY

2-35. The OSAT depends on the medical and theater command and the units with which they work for transportation. The OSAT personnel are issued personal weapons and protective equipment, but depend on the units they are with for protection.

COMMUNICATIONS

2-36. The OSAT depends on the medical and theater command and the units with which they work for communications. With approval of theater command, they may be allowed USAMRMC-issued phones for satellite communication for mission-essential purposes.

ORGANIZATION

2-37. The typical OSAT has three officers and two enlisted Soldiers. The officer positions are listed as research psychologists (AOC 71H), but may be filled by clinically qualified BH officers with research training and experience. The enlisted are usually MH specialists (MOS 68X) with research training and experience. Staffing may also include personnel with information management specialties. The equipment is the required personal equipment of the team members, plus data collection and automated data processing equipment appropriate to the mission.

Chapter 3

Combat and Operational Stress Control Support Operations

SECTION I — COMBAT AND OPERATIONAL STRESS CONTROL PROFESSIONAL DISCIPLINES AND PROFESSIONAL CONSULTANTS

PROFESSIONAL DISCIPLINES

3-1. There are five BH professional disciplines and two enlisted specialties that serve in support of the COSC mission. The professional disciplines and enlisted specialties were identified in Chapter 2. The professional disciplines include social work, clinical psychology, psychiatry, occupational therapy (OT), and psychiatric/BH nursing; the enlisted specialties are in BH and OT. While much of the COSC knowledge base and most of the skills are shared by all the BH personnel, each discipline brings its own perspective from its professional training, skills that can only partially be familiarized to the others, and in some cases unique credentials to conduct specific assessments and treatments. The special perspectives and skill sets of the officer and enlisted members of the COSC team are summarized in the following subparagraphs.

OFFICER AREAS OF CONCENTRATION

Social Work Officer

3-2. The social work officer (AOC 73A67) evaluates the psychosocial systems affecting the Soldier, his family, and the unit; identifies and coordinates with military and civilian support agencies (including the Army Substance Abuse Program and the Family Advocacy Program [FAP]); assesses BH disorders; provides a range of counseling and psychotherapies. Installation MTFs are responsible for BH services that are beyond the scope of COSC functional areas such as long-term clinical care, inpatient services, and alcohol treatment programs. This officer provides proactive consultation, conducts individual and group counseling, supervises Soldier restoration/reconditioning, and coordinates RTD of recovered cases. He also provides staff advice and coordinates Army and civilian social services support.

3-3. When deployed as a member of a COSC preventive team or task-organized COSC element, the social work officer's duties include—

- Evaluating psychosocial (unit and family) functioning of Soldiers with COSR and misconduct stress behavior.
- Coordinating and ensuring the return of recovered COSR and neuropsychiatric (NP) Soldiers to duty and their reintegration into their original or new unit.
- Identifying and resolving organizational and social environmental factors which interfere with combat readiness.
- Coordinating support for Soldiers and their families through Army and civilian community support agencies, when possible.
- Apprising unit leaders, primary care physicians, and others health care providers of available social service resources.
- Providing consultation to supported unit commanders and to other BH/COSC personnel regarding problem cases.
- Counseling and providing therapy for Soldiers with psychological problems.
- Conducting and supervising unit needs assessment and TEM.

3-4. The doctoral level social work officer, as outlined in DODD 6490.1, may conduct clinical evaluations for imminent risk or dangerousness.

Clinical Psychologist

3-5. A clinical psychologist (AOC 73B67) diagnoses BH disorders and a range of psychotherapies including cognitive/mental approaches, assessment for safety evaluations, and psychometric assessment.

Psychiatrist

3-6. The psychiatrist (AOC 60W00) is a physician that is credentialed to diagnose and treat BDPs. He also conducts mental assessments, makes differential diagnosis, provides a range of psychotherapies and safety evaluations, and prescribes medication.

Psychiatric/Behavorial Health Nurse

3-7. This AN officer (AOC 66C007T) provides specialized nursing services for emotionally distressed individuals and promotes COSC within the MTF and the adjacent military community. He performs liaison and consultative functions to ensure continuity of patient care. The additional skill identifier of 7T indicates that advanced skills and competencies are required for this position and that the individual may act as a consultant in clinical nursing practices.

Occupational Therapist

3-8. The occupational therapist (AOC 65A00), plans, implements, and supervises preventive and restorative OT services to enhance the occupational performance of Soldiers. This officer evaluates Soldier's performance across the spectrum of occupational areas (such as activities of daily living, work, education, leisure, and social participation). Analyze jobs and job tasks for underlying performance requisites. Identify and evaluate mental and physical stressors and stress reactions and teach prevention, adaptive coping and psychosocial skills. Apply therapeutic media in both, individual and group settings to include environmental adaptation, the therapeutic use of self, consultation, education, and occupation (purposeful activity). Match Soldiers to therapeutic and/or vocational activities based on therapeutic need and functional ability.

ENLISTED MEDICAL OCCUPATIONAL SPECIALTIES

Mental Health Specialist

3-9. The MH specialist (MOS 68X30/20) provides the perspective of the enlisted Soldier, takes the initial history and conducts mental status evaluations for supervising officer, and administers psychological tests. The MH specialist also performs counseling and may have advanced specialty training (such as drug and alcohol counseling with higher ranked individuals).

Occupational Therapy Specialist

3-10. The OT specialist (MOS 68W30/20N3) provides the perspective of the enlisted Soldier, assists the supervising OT officer in evaluating a Soldier's occupational performance; conducts initial occupational performance history interviews and mental status evaluations; observes clients to gather data as part of task performance skill assessments; and implements OT interventions under the supervision of an OT officer.

CONSULTANTS

3-11. There are two levels of COSC consultants—the theater-level consultant and the subordinate commander consultant. The theater-level COSC consultant is normally the senior COSC officer designated by the theater surgeon and approved by the theater commander. The subordinate command COSC consultants are MH officers that function as the point of contact (POC) for one or more commanders.

These consultants oversee, coordinate, and conduct COSC interventions and activities for their respective commanders. They accomplish the following tasks appropriate to their level of operation:

- Direct and report AO-wide COSC needs assessments.
- Coordinate AO-wide implementation of COSC functional areas.
- Ensure COSC standards are established and consistently applied.
- Ensure compatibility between COSC standards and hospital BH standards (theater/AO COSC consultant).
- Define COSC operational requirements.
- Recommend the AO COSC evacuation policy (theater/AO COSC consultant).
- Prepare the COSC portion of the FHP estimate and operations plan.
- Review all command medical policies affecting COSC interventions and activities.
- Evaluate the quality of COSC interventions and activities rendered in the AO.
- Make regular AO COSC consultant site visits to COSC sections and units.
- Coordinate with other COSC consultants in adjoining AOs and in the primary medical support elements to ensure unity of effort.
- Oversee AO COSC training.
- Coordinate joint and combined COSC interventions and activities as necessary.
- Plan for future COSC operations.
- Consult with the command surgeon on all COSC matters.

SECTION II — DEPLOYMENT AND EMPLOYMENT OF MENTAL HEALTH SECTIONS

MENTAL HEALTH SECTIONS

3-12. The MH section assignments, staff, functions and responsibilities were discussed in Chapter 2. The information provided in this section pertains to deployment and employment and how these sections prepare for COSC support for their units in garrison, combat operations, and stability and reconstruction operations predeployment. The involvement of MH section with support units during field and other training exercises is important. It permits BH personnel to gain a familiarity with type of units supported, how they conduct operations, their mission, and the likely stressors associated with a particular type of unit. It also allows them to interact and become familiar with unit personnel and to gain some level of trust through their interactions. Combat and operational stress control personnel should be active participants in all training exercises. Unit predeployment training must be planned when the unit is alerted for its deployment but scheduled after cross-leveling and filling of all positions is completed. Training of BH personnel will include—

- Ensuring their active participation in the normal training cycle of the BCT/ACR to which assigned.
- Training the BH officers and specialists in all COSC functional areas that support their unit.
- Working with the supported units on a frequent basis so BH personnel may gain a familiarity with the unit's personnel, capabilities, and possible stressors associated with its mission.
- Providing predeployment train-up to include refresher and operation-specific training before deployment.

PLANNING

3-13. Since planning is normally performed at battalion level, MH sections are involved in developing a COSC estimate to support the unit plan. The COSC estimate of the situation is derived from mission analysis, to include considerations for COSR, BH disorders, substance abuse, misconduct behaviors, suicide, major stressors, troop populations to be supported, and COSC augmentation requirements or assets available.

Coordinating

3-14. The BH personnel may be involved in coordinating COSC support with supporting and supported units, as appropriate. Brigade BH personnel must be known and trusted by the leaders of the units they support. Brigade BH personnel coordinate with other BH officer and enlisted personnel pertaining to the professional disciplines associated with conducting the COSC mission. They also coordinate with other key medical and nonmedical personnel (such as health care providers, surgeons, chaplains, UMTs that assist or may be involved with the COSC mission).

Deployment of Mental Health Sections

Early Deployment

3-15. The MH section should be deployed early into staging areas with the lead elements of the medical company and supported units.

Command and Control

3-16. The MH section falls under the C2 of the FSMC/BSMC and may receive mission guidance from the brigade surgeon through the FSMC/BSMC commander. Tasking for COSC operational support flows through the BSMC/FSMC headquarters. When BH personnel from the medical detachment/company, CSC are attached, the MH section ensures the timely flow of operations and situation awareness information to that element.

Employment of the Mental Health Sections

3-17. Employment of the MH section is based on the type of operations being conducted. The MH section normally operates from the FSMC/BSMC located in the brigade support area (BSA). When MH personnel are deployed forward, they are dependent on supported units for rations and logistical support.

Mental Health Functional Area Considerations

3-18. When determining COSC support for BCT units, the BH team should review each of the BH functional areas and determine which BH functions need to be implemented for each supported unit. Also, in developing the COSC plan, the BH team should consider—

- Conducting a UNA for each supported battalion and separate company supported.
- Establishing a consultation and education/training program for supported units to include—
  - Traumatic event and transition management.
  - The need for COSC triage, stabilization, and BH treatment should be addressed as they arise.
  - Soldier restoration, which can only be accomplished within and by the BSMC holding section or with augmentation by or referral to CSC unit personnel.
  - Reconditioning services that will need to be referred to nearest location where available.

SECTION III — SUPPORT OPERATIONS CONDUCTED BY THE MEDICAL UNIT, COMBAT STRESS CONTROL

PREDEPLOYMENT ACTIVITIES FOR ALL MEDICAL UNITS, COMBAT STRESS CONTROL

3-19. The predeployment activities identified in Section II for MH sections also apply for the units in Section III that include the ASMB or MMB and medical companies and detachments, CSC. These units must also keep themselves focused on applying the nine functional areas during the predeployment phase.

Note. The MMB is the C2 organization which will replace the ASMB, the medical logistics battalion, and the evacuation battalion.

Medical Detachment, Combat Stress Control Deployment

3-20. The medical detachment, CSC is deployed to provide COSC support whenever a division or two BCTs/regiments are deployed. It is assigned to a medical brigade or other medical headquarters for C2. Elements of the detachment may be further attached to FSMCs/BSMCs for operational control and logistical support while in support of the BCT.

Medical Company, Combat Stress Control Deployment

3-21. The primary mission of this unit is to provide COSC support throughout a corps AO. It is assigned to either a MEDCOM/MEDBDE and may support from two or up to five divisions. They assign teams to any operational area based on troop populations without BH support. They move teams rapidly to areas of surge workload or mass casualties.

Medical Force 2000 Verses the Medical Reengineering Initiative Detachments

3-22. The Medical Force 2000 (MF2K) and MRI CSC detachments conduct COSC support operations very similarly. However, in the MRI detachment, psychiatry assets are assigned to the fitness section and psychology assets are assigned forward to the prevention section as compared to the MF2K detachment, the manning strategy for psychology and psychiatry is reversed. The psychologists are assigned to the fitness section and psychiatrists are assigned to the prevention teams.

Medical Detachment, Combat Stress Control Support to a Division and Brigade Combat Teams During Combat Operations

3-23. When in support of an AOE division, a medical detachment, CSC is usually attached to the MSMC/BSMC and for a transformation division the medical detachment, CSC is attached to the MMB or MEDBDE. It can be attached to other battalion-sized units for support. It is under the operational control of the support battalion and MSMC/FSMC/BSMC and closely coordinates its augmentation support with the division surgeon and division psychiatrist. Habitual long-term relationships of medical detachments, CSC with specific divisions should be established so that COSC support is integrated as a normal function of the FHP mission. However, as an echelon above division (EAD) asset, the detachment (or its modular teams) may be task-organized and deployed to support other units or missions other than divisional units, as workloads require. The medical detachment, CSC depends on the units to which it is attached for administrative and logistics support. It may be reinforced and/or receive personnel replacements from a medical company, CSC or from another medical detachment, CSC.

Preventive Section Operations

3-24. Upon arrival to a theater, the detachment is initially attached to a higher medical headquarters for C2 and its three CSCP teams of four-person each are usually further attached. One CSCP is attached to each divisional MSMC prior to commencement of combat operations. This will allow the teams to linkup with and augment division COSC assets in support of the brigades.

- The deployment of each CSCP provides additional COSC personnel that increase COSC augmentation capabilities at the FSMC. A CSCP may routinely visit the BASs to provide consultation. The CSCP may deploy, as needed, to provide reconstitution support to units undergoing hasty or deliberate reorganization.
- In some operations, some or all of the CSCP personnel may either be attached to an ASMC or are deployed in support of a maneuver brigade or reconstitution site. It is important for all CSCP personnel to understand that they are responsible for all functional areas, and will conduct them as dictated by the mission.

Fitness Section Operations

3-25. The CSCF may locate with any Level II or above MTF in the AO. This section has tents and is equipped to operate a Soldier restoration center and can supervise Soldier restoration activities to include consultation, rest, relaxation, and treatment as necessary for speedy RTD within three days. The CSCF also provides triage and stabilization at the MSMC and consultation to other division units located in the DSA and division rear areas. All CSCF members are responsible for providing support for all COSC functional areas as dictated by the mission.

3-26. The CSCF staff may move within the AO to temporarily reinforce or reconstitute a CSCP at the brigade level, or to escort COSR casualties to the Soldier restoration center. The CSCF personnel give reconstitution support to attrited units, especially when the units return to the DSA. The CSCF supplements the division psychiatrist; brigade COSC teams, CSCPs, and chaplains and leaders in after-action debriefings. The CSCF helps integrate recovered Soldiers and new replacements into units during reorganization activities. When either the BSA or DSA is tactically too unstable to allow Soldier restoration operations, the CSCF may locate with an ASMC or a CSH and attempt to continue their support to the division from that location.

COMBAT STRESS CONTROL DETACHMENT IN CONTINGENCY OPERATIONS

3-27. For contingency operations, a task-organized CSC element from the CSC medical detachment may be deployed. A CSCP supporting a maneuver brigade may operate out of a central base of operations. The team may deploy forward to the BSA (base camps or fire bases), in response to—

- Anticipated battle.
- After-action debriefing requirements.
- Alcohol/drug problems in a unit.
- Incidents of misconduct stress behaviors.
- High incidence of COSRs.
- Unit rotation in or out of theater.

3-28. Several CSCF and CSCP from two or more CSC medical detachments may be consolidated under the command of the senior medical detachment, CSC commander to staff a central reconditioning program for the AO. This may also function as an alcohol/drug detoxification rehabilitation program. This may be referred to as a COSC center and will also provide consultation and treatment support to MP confinement facilities where misconduct stress behaviors may have led to incarceration. This could also include support for EPW and detainee confinement facilities.

STABILITY AND RECONSTRUCTION OPERATIONS

3-29. In stability and reconstruction operations, if the force deployed is smaller than a division, a medical detachment, CSC would not be required to provide COSC support. In such cases, the medical company, CSC or the medical detachment, CSC may be tasked with providing either a COSC team or a task-organized COSC element. The COSC team or the task-organized COSC element is attached to the supporting medical headquarters or to an MTF and conducts its mobile consultation mission.

MEDICAL UNITS, COMBAT STRESS CONTROL INTERFACE AND COORDINATION REQUIREMENTS

3-30. The medical units CSC must interface with its higher headquarters element and with the unit to which it is attached. The higher headquarters may be a MEDCOM, MEDBDE, or an ASMB/MMB.

- Interface between the CSC units and their higher headquarters. The medical unit CSC interfaces with its higher headquarters pertaining to its assigned mission. It provides estimates and has input to the operation plan (OPLAN). The unit receives its OPORD from the higher headquarters. Interface between the unit and higher headquarters staff elements will include the following subject areas—

- The COSC operations.
- Assignment or attachment of the medical detachment, CSC elements.
- Daily personnel and equipment status reports.
- Class VIII (medical supply) status and resupply requirements.
- Casualty feeder reports.
- Operation orders.
- Personnel replacement for the detachment.
- Medical intelligence information.
- Behavioral health/COSC consultation tasking and results.
- Maintenance requirements and requests.
- Replacement and reconstitution operations.
- Civil-military operations.
- Communications (signal operation instructions [SOI], access to message centers and nets, and transmission of COSC messages through medical, land, and other channels).
- Mass casualty plan.
- Road movement clearances.
- Tactical updates.
- Contingency operations.
- Return-to-duty and non-RTD procedures.
- Medical evacuation procedures (air and ground ambulances).
- Changes in locations of supported unit.

- Interface and coordination with unit to which attached. The headquarters of the unit to which a medical detachment, CSC is under operational control (OPCON) or attached is responsible for providing the administrative and logistical support requirements of the detachment. These requirements are normally identified in the attachment order. The higher headquarters will be identified in the attachment order and the medical detachment, CSC will be coordinated with this headquarters prior to deployment. The medical detachment, CSC must coordinate with the headquarters staff according to the tactical standing operating procedures (TSOP) of the unit of attachment. The staff shares information with the detachment commander or his representative pertaining to the threat, tactical situation, patient/COSR casualty status, and changes in FHP requirements. Coordination activities and subject area information exchange should include—
  - Command and control procedures.
  - Status of FSMCs/BSMCs and CSCPs.
  - Communications and SOI.
  - Operational support requirements.
  - Civil-military operations.
  - Soldier restoration operations.
  - Reinforcement and personnel replacement.
  - Road movement and clearances.
  - Casualty reporting and accountability.
  - Patient-holding procedures.
  - Force protection.
  - Convoy operations.
  - Geneva Conventions.
  - Detainees/EPW.
  - Improvised explosive devices (IEDs).
  - Religious and cultural considerations.

RECONDITIONING CENTERS

3-31. The medical company, CSC task-organizes CSC elements to staff separate small reconditioning centers in locations that are relatively secure. However, under some circumstances, the company may consolidate teams to establish a large reconditioning center, which supports two or three divisions. Reconditioning facilities normally locate near a CSH. See Chapter 12 for definitive information on reconditioning center operations.

Chapter 4

Unit Needs Assessment

SECTION I — PRINCIPLES AND TENETS OF UNIT NEEDS ASSESSMENT

GENERAL PRINCIPLES

4-1. The UNA is a systematic process for identifying the COSC needs of units. The UNA allows COSC personnel to identify priorities for interventions and activities and for allocating resources. The UNA is not a clinical screening to identify individuals who have or are at risk for BH disorders problems, but rather evaluates the needs of the Soldier population and leads to more effective preventive COSC activities and early interventions. The UNA allows COSC personnel to—

- Identify and describe specific areas of COSC need.
- Discover factors contributing to the needs.
- Provide an assessment of the BH training needs of Soldiers, leaders, UMT, and medical personnel within the unit.
- Develop plans to meet or improve the COSC needs of Soldiers and units through prevention and early intervention activities.

SCOPE

4-2. Unit needs assessments can be conducted at various command levels from small to large units. The COSC personnel at each level should conduct UNAs for their supported units. Larger-unit UNAs can include the composite findings and recommendations of one or more UNAs completed at subordinate levels. For example, a brigade UNA may include the results of several subordinate battalion UNAs (consolidated for confidentiality). Generally, UNAs are not conducted below the company level, though exceptional circumstances may dictate a platoon or lower UNA.

TENETS

4-3. Tenets of UNAs—

- Gain commander approval and support prior to conducting the UNA.
- Protect anonymity and ensure confidentiality of Soldiers and commanders. This includes the protection of unit identification from higher headquarters.
- Provide the commander with an unbiased assessment.
- Consider the social, political, and organizational factors of the environment.
- Ensure that information sources represent the entire unit.
- Select an assessment method that is consistent with the operational situation.
- Limit overgeneralizing the findings from one unit to another or from one time or situation to another.
- Recognize that UNAs provide population-level assessments of COSC needs, not clinical screening tools to identify individuals who may benefit from COSC interventions.
- Distinguish between what respondents report they need and what interventions are required.
- Ensure the UNA is planned and coordinated at a level commensurate with the complexity of the assessment and/or situation before starting.

SECTION II — FOCUS AND METHODS OF DETERMINING UNIT NEEDS ASSESSMENT

AREAS OF FOCUS FOR A UNIT NEEDS ASSESSMENT

4-4. A UNA involves the systematic assessment of numerous areas of Soldier and unit functioning. A typical UNA includes, but is not limited to, the following areas:

- Major stressors impacting the unit.
- Level of unit cohesion.
- Well-being of unit Soldiers.
- Soldier concern about home-front issues.
- Soldier knowledge and skill for controlling combat and operational stress.
- Soldier ideas for addressing COSC needs.
- Soldier knowledge of accessing COSC resources.
- Barriers and stigma that prevent Soldiers from accessing COSC services.
- Training needs of Soldiers, leaders, UMT, and medical personnel on topics of COSC importance (such as buddy aid, suicide awareness, or suicide prevention). See FM 6-22.5, FM 22-51, and US Army Training and Doctrine Command (TRADOC) Pamphlet 600-22 (available at: http://www.tradoc.army.mil/tpubs/pamndx.htm) for additional information on suicide awareness and prevention.

METHODS OF UNIT NEEDS ASSESSMENTS

4-5. The UNA takes advantage of all available information. Various methods can be used to assess general unit needs and to identify issues that differ among subgroups (for example, gender, rank, or race/ethnicity). The use of multiple assessment methods is recommended. These methods may include—

- Interviewing Soldiers to hear their perceptions and concerns. Information gathered during the casual conversation, although informal, may afford valuable anecdotal information (as well as developing trust and familiarity).
- Reviewing policy documents (standing operating procedures [SOPs]).
- Interviewing key unit personnel (chain of command, chaplain, and medical personnel).
- Conducting structured group interviews (focus groups or unit survey interviews).
- Administering standardized surveys and questionnaires (paper/pencil or web-based).
- Monitoring trend indicators (such as high rates of BH referrals, sick call, or misconduct, Soldier suicide, sexual assault, fratricide or disciplinary actions).
- Using multiple methods for gathering information when possible to ensure different viewpoints are considered.

Planning Considerations for Unit Needs Assessment

4-6. The UNA varies in complexity and formality depending on the purpose of the assessment and the needs of the supported unit commander. Many factors determine the complexity and formality of a UNA, which in turn influence its feasibility.

Complexity

4-7. Complexity equates to the requirements and cost to complete a UNA. Complexity is influenced by factors such as—

- The size and number of units to be assessed.
- Geographic dispersion of the units and time constraints.

Formality

4-8. Formality equates to the degree that scientific principles and methods are employed in the conduct of the UNA. An example of a UNA with low formality is one where the assessing team uses nonstructured group and individual interviews and perhaps brief questionnaires they have developed. The information is obtained from key individuals and a convenience sample of troops. That may be sufficient to quickly identify problems and make recommendations to resolve them. A UNA with high formality is one that uses professionally validated questionnaires and structured focus group interviews according to standard protocols. The data is collected from a scientifically selected sample of the larger population. The data is analyzed by standard analysis programs, so that statistically selected samples of valid comparisons can be made with other similarly sampled units in other geographical locations and across time. The OSAT expertise and assistance may be necessary for some UNA instruments and methods. In general, increasing formality increases the complexity of the UNA. The UNA requires a higher level of formality as the need for objective data and scientific precision increases (such as when results from multiple units are to be merged or compared with other UNAs).

Feasibility

4-9. Feasibility is the ability to accomplish a UNA with available resources. In developing a particular UNA, trade-offs are made to achieve an acceptable level of data quality (formality) for an affordable cost (complexity).

UNIT NEEDS ASSESSMENT PROCESS

4-10. All UNAs are conducted following a three phase plan.

PREASSESSMENT

4-11. Preassessment is an initial phase to obtain command support, determine target issues, and select appropriate methods to use.

ASSESSMENT

4-12. Assessment is the phase for gathering, integrating, and analyzing information to identify the COSC needs of the unit.

POSTASSESSMENT

4-13. The principal task of the postassessment phase is to determine the COAs to present to the commander which address the identified COSC needs. These findings are then linked to a plan of action.

WHEN TO CONDUCT UNIT NEEDS ASSESSMENTS

4-14. Unit needs assessments may be conducted at various times throughout the deployment cycle. A UNA may be conducted—

- Prior to initiation of COSC interventions and activities while COSC personnel and units establish their support relationships.
- To assess the effectiveness of COSC interventions and activities that are in the process of being conducted or that have been completed.
- At the request of supported commanders.
- To monitor trend indicators (see Paragraph 4-5).
- After serious traumatic events and significant unit transitions.
- To collect unit information for COSC planning.

4-15. Unit needs assessments should be conducted during all types of deployments, including stability and reconstruction operations, and combat operations.

Chapter 5

Consultation and Education

SECTION I — PRINCIPLES AND TENETS OF CONSULTATION AND EDUCATION

GENERAL PRINCIPLES

5-1. Combat and operational stress control consultation is defined as the transmission of information through an interactive relationship between the consultant and consultees. Education is used here in a broader sense in that it is the transmission of information by any means. Examples of consultation include providing COSC advice, coaching, training, and planning assistance. During this process, the consultant learns about the consultees and their needs, and tailors the interactions accordingly. Examples of education include distributing flyers, video and radio broadcasts, and news articles. The consultation and education functional area supports the other COSC functional areas.

CONSULTANTS

5-2. The term BH consultant is used in the general sense to describe any person performing COSC consultation or education. In addition to those theater or subordinate commander COSC consultants identified in Chapter 3, all BH/COSC personnel may serve as consultants at their level. Familiarization training with other BH disciplines enriches the ability to serve as a consultant.

CONSULTEES

5-3. Depending on the units in the area of support, broad types or categories of personnel may be consultees. These personnel are involved in recognition and control of stress as a result of their position or duty assignments. Consultation and education consultees may include—

- The command or unit surgeon and his staff.
- Staff chaplain and UMT.
- The senior commander and the senior NCO of a battalion, brigade, division, or corps.
- Staff officers and NCOs, including adjutant and personnel (S1/G1), intelligence (S2/G2), operations (S3/G3), civil-military affairs (G9), and the Judge Advocate General (JAG).
- Medical personnel such as PVNTMED teams that have missions that often complement the COSC mission.
- Company grade leaders, especially company commanders, executive officers, first sergeants (1SGs) platoon leaders, platoon sergeants, company NCOs, and Soldiers that are trained to be peer mentors.

SOLDIER-PEER MENTORS

5-4. Soldiers selected by their commanders may be trained to provide COSC help-in-place assistance for COSC information to peers. They may also serve as a POC between fellow Soldiers and the COSC and UMT teams. Selected Soldier-peer mentors with additional training could assist commanders with conducting COSC training in mission risk assessments.

TENETS FOR CONSULTATION AND EDUCATION

5-5. Consultation and education are ongoing processes that are performed across the deployment cycle and the continuum of operations. They may be provided in response to a specific request by command or recommendation of COSC personnel. Consultation and education may be provided during routine scheduled meetings such as a commander's weekly update or a leader professional development class.

5-6. Effective consultation is accomplished by the consultant's active outreach. Consultation is best conducted through recurring face-to-face contact, preferably at the consultee's location. Telephone and radio may be used to setup initial meetings and provide follow-up consultation. Consultation may be conducted one-on-one or in small groups where interaction is feasible. When necessary and feasible, audio or video teleconference may suffice. Active outreach supports the functional areas particularly COSC triage and traumatic event and transition management.

5-7. Successful consultation depends on the consultant's credibility, and the trust and familiarity established with consultees. In addition to sound professional knowledge base and clinical skills, the consultant must have military bearing and knowledge of the military (including the units, missions, vocabulary, acronyms, and skills involved). Rapport is enhanced by the demonstration of the consultant's genuine interest in the consultee and the unit. There are some situations when consultation and education are more effectively provided by the COSC officers, the NCOs, or the junior enlisted personnel.

SECTION II — CONSULTATION, EDUCATION, AND PLANNING

CONSULTATION PROCESS

5-8. The following six steps outline the process for most consultations:

- Initiate the process by introducing yourself and your capabilities to key leadership and get approval to continue.
- Assess the needs of consultees and formulate ways to address them.
- Present COA to consultee and define goals and feasibility of alternative actions.
- Implement approved COA.
- Evaluate outcomes or progress.
- Plan follow-up actions.

5-9. Consultation and education encompasses a broad range of topics, extending from prevention to treatment. The process of consultation and education assists the consultee with anticipation, identification, and control of stressors (environmental, physiological, cognitive, and emotional) and stress reactions (adaptive and maladaptive stress reactions). The following list of COSC functions, resources and their availability provides examples of consultation and education topics:

- Operation risk factors.
- Individual risk factors.
- Stress moderators.
- Unit cohesion, esprit de corps, and morale.
- Critical event management.
- Barriers to care and overcoming stigma.
- Operations planning.
- Integration of new personnel to unit.
- Trust in equipment and supporting units.
- Tough realistic training.
- Home-front stressors.
- Indicators of unit stress level.
- Combat and operational stress behaviors.
- Combat and operational stress reactions.

- Misconduct stress behaviors.
- Mental disorders.
- Suicidal and homicidal behaviors.
- Command referral processes.
- Scapegoating behaviors.
- Rumor control.
- Physical needs (such as sleep, nutrition, and hygiene).
- Social/emotional needs (such as morale, welfare, recreation push package).
- Communication with home.
- Reintegration of Soldiers into unit.
- Alcohol and substance abuse.
- Procedures for COSC triage.
- Care provider stress (such as compassion fatigue, vicarious trauma, moral dilemma).
- Long-term consequences of COSR and mental disorders.
- Medical evacuation policy and procedures.
- Leadership.
- Fear management.
- Rest and relaxation (R&R) considerations.

5-10. Additional details about some of these topics may be found in FM 22-51.

5-11. From theater staff level down to the MH sections, BH personnel are involved in the consultation and planning process for COSC support. Their support may be either through providing situational updates, developing staff estimates, or being directly involved and assisting with the development of the FHP plan. Planning starts with mission analysis. The mission analysis is the first step of the military decision making in abbreviated planning for a time-constrained situation, see FM 5-0. A part of mission analysis is based on the commander's intent and guidance. A medical battalion headquarters staff may receive addition instructions from the MEDBDE headquarters staff. During the planning process, command surgeons and medical planners (S3/G3) may seek information from the BH/COSC consultant or task medical company/detachment, CSC for COSC estimates.

MISSION ANALYSIS

5-12. The COSC planning is an essential part of every medical estimate and operations plan. Ensure that the command surgeon involves the COSC consultant in all medical planning. Mission analysis includes—

- Assessing COSC capabilities (organic and attached assets with current status and location).
- Gathering information from the commander's guidance regarding the upcoming operations. (See the S2 for intelligence, the S3 for scheme of maneuver and forces involved; the S1 for casualty estimates, personnel status issues, and units supported, and the logistics staff officer [S4/G4] for logistical considerations). Surgeons provide disease and nonbattle injuries (DNBI) estimates and status of medical units. Unit needs assessments, organic medical personnel, and UMT can provide additional information.
- Developing the COSC estimate based on the gathered information. The COSC estimate helps medical planners to anticipate demands on medical and COSC resources and to prepare accordingly. See FM 3-0, FM 5-0, and FM 8-55 for methods to develop the COSC estimate.
- Assessing limitations (specify reasons that BH assets are not available).
- Identifying specified, implied and essential COSC task in the FHP annex of the OPORD.
- Providing COSC COA.
- Presenting estimate and COA to command surgeon. Define goals and feasibility.
- Implementing approved COA. Consult with implementing COSC personnel about estimate and operational details.

- Evaluating outcomes or progress.
- Providing estimate and COA for fragmentary order (FRAGOs) and planning follow-up actions.

REHEARSAL

5-13. To achieve optimal synchronization, the FHP plan is rehearsed as an integral part of the combined arms plan at the combined arms rehearsal. Medical leaders provide input to the over all plan and develop the concept for the FHP plan. During the decision-making/orders process, they identify critical events and synchronize the FHP plans. In addition to medical locations on the CSS overlay, these plans indicate the triggers for FHP events. At the rehearsals, leaders practice their synchronized plans that include FHP.

5-14. The CSS/FHP annex of the OPORD that includes map overlays is the conclusion of the medical planning efforts; the rehearsal is the culmination of the preparation phase for an operation. The medical platoon leader has the responsibility for rehearsing FHP operations. Rehearsals are done to achieve a common understanding and a picture of how the plan will be implemented. The following leads to a successful rehearsal—

- All plans must be completed prior to the rehearsal.
- The FHP portion of the battalion rehearsals should focus on the events that are critical to mission accomplishment. A successful rehearsal ensures explicit understanding by subordinate medical personnel of their individual missions; how their missions relate to each other; and how each mission relates to the commander's plan. It is important for all medical units to understand the complete FHP concept.
- Rehearsing key FHP actions allows participants to become familiar with the operation and to visualize the "triggers" which identify the circumstances and timing for friendly actions. This visual impression helps them understand both their environment and their relationship to other units during the operation. The repetition of critical medical tasks during the rehearsal helps leaders remember the sequence of key actions within the operation and when they are executed.
- The OPORD is issued through effective troop leading procedures.

SECTION III — TRANSITION MANAGEMENT AND SUPPORT IN THE DEPLOYMENT CYCLE

SUPPORTING TRANSITION AND PHASES OF A DEPLOYMENT

5-15. Transition and phases of a deployment cycle may have different and sometimes unique stressors. These stressors may be subjects for consultation and education. Needs assessment may be necessary to redistribute COSC assets and change the means for delivering COSC activities and interventions.

5-16. The COSC personnel conduct transition (change of command) workshops, especially after relief of commanders. These workshops are normally requested by the incoming commander. The purpose of these workshops is to—

- Facilitate staff discussion of what the staff sees as the unit and staff's strong points and the areas needing more work.
- Provide the new commander the opportunity to discuss his leadership style and his expectations and set priorities for the staff.

5-17. The change of OPTEMP may include—

- Going from deployment to combat (offensive or defensive operations).
- Conducting battle handover in place by turning the battle over to another unit.
- Going from high intensity combat to stability and reconstruction operations.

PREDEPLOYMENT SUPPORT

5-18. Medical support in predeployment is mainly the responsibility of MEDDAC personnel or of a US Army Reserve (USAR) mobilization site augmentation unit (which has no BH personnel assigned).

Division and brigade COSC BH personnel are involved with their unit's Soldiers who are being treated for mental disorders, and with general COSC prevention of the unit members and families. The CSC units that are not deploying may be tasked to support predeployment COSC activities and interventions. Combat and operational stress control screening of individuals during predeployment may include—

- Secondary BH screening for Soldiers who are referred by primary care providers that flagged positive on DD Form 2795 (Pre-Deployment Health Assessment).
- Health screening of medical and BH records.
- Fitness for duty evaluations according to AR 40-501.
- Evaluations and recommendations to leadership about Soldiers who are fit for duty overall, but should not be deployed at this phase of the operation for BH reasons.

5-19. Other predeployment functions may be provided on request. These functions may include—

- Briefings, consultation and education on deployment cycle stressors and how to cope, and on specific stressors that Soldiers may encounter in the AO.
- Unit assessments at request of commanders.

MID-TOUR REST AND RECUPERATION OR EMERGENCY LEAVE

5-20. Leaders must be aware and alert for Soldiers who exhibit a need for COSC screening prior to mid-tour R&R. Other Soldiers may only require some education briefings and handout cards on handling their transition from a hostile CZ to R&R environments. In some cases, prior to emergency leave, some screening for domestic violence risk may also be necessary in any potential high-risk Soldier. This COSC screening is done for ensuring the safety of the Soldier and of others. One protocol has unit commanders referring those Soldiers that have requested leave for the purpose of addressing marital, legal, or other highly charged difficulties at home, to the BH officer. The BH officer explores the situation and the Soldier's reactions to it and checks for history with the rear detachment of the MH section, MEDDAC/MEDCEN and FAP. If the BH officer finds the Soldier to be at high risk, the commander can deny leave. Another option is to escort the Soldier to home base. At home base, he is kept by his unit's rear detachment until a meeting is held under safe conditions with the Soldier, BH, FAP personnel, his family, and the others involved in the difficulty. At this meeting the situation is defused or resolved before the Soldier is permitted to leave the rear detachment area.

REDEPLOYMENT

5-21. For Soldiers, redeployment is the process of getting orders to redeploy, embarking towards the demobilization site, or deploying to another out-of-the-theater mission. Screening of these individual Soldiers may be a requirement and have established criteria that should be used to ensure appropriate screening is conducted. The theater or specific units may require that all primary care providers who perform screening interview, postdeployment health assessment have a designated BH consultant. Available COSC/BH personnel provide secondary one-on-one BH screening for Soldiers who are referred by primary care providers because they flagged positive on DD Form 2795. Some Soldiers may require brief treatment or referral may be required as a result of the screening.

END-OF-TOUR STRESS MANAGEMENT

5-22. End-of-tour stress management is essential to reduce mental problems with Soldiers returning to their home station and/or families. This also promotes Soldiers seeking help early when problems occur. Mental health/COSC personnel consult with command on activities during redeployment and for postredeployment. Issues may include—

- Common mental disorders and potential misconduct may result when Soldiers are relieved of the stress and the focus of dangerous missions and are looking towards and preparing to go home.
- Memorial services for those Soldiers who died during the deployment.
- Recognition of outstanding performance and the equitable awarding of decorations.
- Closure ceremonies.
- Homecoming-reunion education briefing and training.

5-23. The end-of-tour debriefings (EOTDs) are like leader-led after-action debriefings, except that the time period covered extends from predeployment through to this point in redeployment. The EOTD should be conducted while the unit is still in the theater or at some place where transportation home pauses for one or more days. For each phase of the operation, the participants review what were the significant events and the problems or stressor that bothered the group (including critical events, if any). Equally important the group recalls what went well, how problems and stressors were overcome (at least by some members), and what positive memories and feelings they had and will take home.

Facilitators

5-24. The COSC or UMT facilitators should be invited to advise the leader and sit in if there are likely to be many negative memories and feelings that have disrupted or threaten unit cohesion. Advise leaders and troops on the importance of sustaining unit identity and contact with teammates beyond postdeployment and in the future.

Postdeployment

5-25. Postdeployment covers that period of time from embarking from theater, through scheduled postdeployment activities, and return to work after block leave. As with predeployment, medical support in postdeployment is mainly the responsibility of installation MTF, often augmented by a USAR mobilization site augmentation unit. Division and brigade COSC/BH personnel and CSC units that are themselves demobilizing as their units are deactivating will have limited involvement with the BH and COSC needs of fellow Soldiers. These personnel continue to provide, as needed, consultation to the command and to key personnel of the unit that they were assigned/attached. They consult with the installation MTF about BH and coordinate COSC needs that should be met. The CSC units that were not deployed may be tasked to support postdeployment COSC activities and interventions. The COSC personnel who are not demobilizing provide reintegration education. Screening of individuals postdeployment will continue and secondary screening will continue in postdeployment and be recorded on a DD Form 2796 (Post-Deployment Health Assessment). Each primary screener should have a designated BH consultant from the MEDCEN/MEDDAC. Secondary screening and treatment of referrals are provided by the installation MTF. Recent research with a returning division indicated that a secondary postdeployment screening from 90 and 120 days after returning from the theater may be necessary. This seems to be the time when persisting or delayed symptoms, problems, and perhaps impairment show need for intervention. All Soldiers should be advised about follow-up BH support resources. Soldiers that are leaving active service should also be informed about deployment-related entitlements and benefits.

Readiness Phase of the Deployment Cycle

5-26. Division/brigade MH sections and medical units, CSC, of Active Army and Reserve Component (RC) must use their training opportunities to the fullest. During these training exercises, they should provide COSC consultation, education, and training to supported units the same as they would in an actual deployment. Mental health personnel will perform many of their critical mission functions and should not just simulate them. The management of real COSR occurs in field and garrison/home-station settings. Triage and emergency stabilization may be required. Restoration may be appropriate in some field exercises. Even reconditioning can sometimes be provided as a multiday "course" at Active Army posts where CSC units are stationed. Further, TEM and transition management are also potential major roles for organic MH sections and medical units, CSC in peacetime. The RC as well as Active Army medical units, CSC have been used extensively to respond to traumatic events and to assist units that are in the premobilization, mobilization, or postdeployment phases of their deployment cycles.

Chapter 6

Traumatic Event Management

OVERVIEW OF TRAUMATIC EVENT MANAGEMENT

TAILORING TRAUMATIC EVENT MANAGEMENT

6-1. Traumatic event management blends other COSC functional areas to create a flexible set of interventions specifically focused on stress management for units and Soldiers following potentially traumatizing event (PTE). Like other functional areas, COSC providers must tailor COSC and TEM support to the needs of the unit and Soldier.

6-2. An event is considered *potentially traumatic* when it causes individuals or groups to experience intense feelings of terror, horror, helplessness, and/or hopelessness. Guilt, anger, sadness, and dislocation of *world view* or faith are potential emotional/cognitive responses to PTEs. Studies of Soldiers in Operation Iraqi Freedom (OIF) and Operations Enduring Freedom (OEF) have shown a correlation between exposure to combat experiences and BH disorders, most particularly acute stress disorder and PTSD. The following events should be monitored as PTEs for Soldiers and units:

- Heavy or continuous combat operations.
- Death of unit members due to enemy or friendly fire.
- Accidents.
- Serious injury.
- Suicide/homicide.
- Environmental devastation/human suffering.
- Significant home-front issues.
- Operations resulting in the death of civilians or combatants.

Scope of Traumatic Events Management

6-3. In this chapter, TEM is discussed as it applies to military units and personnel to support readiness. Traumatic events management can be adapted to nonmilitary groups or individuals brought together by a PTE. Traumatic events debriefing should be conceptualized as an ongoing process and not an acute intervention. For military units and personnel, TEM is active in all phases of the deployment cycle and across the continuum of military operations.

Traumatic Events Management Functional Mission Area

6-4. Successful TEM relies on a solid foundation of other COSC functional area activities such as UNA that may require the COSC provider to—

- Establish a credible working relationship with supported unit leadership.
- Understand the unit's needs via previously conducted UNAs.

6-5. In the absence of preexisting relationships and UNA, COSC providers should secure command support and recommend a UNA as the first step in TEM. Subsequent UNAs will clarify unit responses to other TEM interventions and to ongoing unit needs.

Unit Needs Assessments

6-6. When conducting a UNA after a PTE the COSC providers need to take under consideration the following—

- Ensure a timely arrival that does not disrupt unit operations but facilitates the UNA.

- They should not interrupt or intrude on the people who are attending to the acute crisis when arriving at the unit, unless asked. Presence without intruding will gain trust information and POC.
- Understanding that the UNA may be limited by the urgency of the unit's return to action, the difficulty of data collection, and having limited resources. Sufficient knowledge must be gained to tailor the interventions to the unit before initiating them. It is better to defer the intervention to the next opportunity if the unit must return to action immediately after replenishment and/or before necessary data is collected.
- Know that the UNA is a unit-level assessment and does not substitute for individual-level screenings or COSC triage. See Chapter 4 for a detailed discussion of UNA.

Consultation and Education

6-7. The COSC provider should conduct unit leader consultation and education activities prior to a PTE. Consultation and education topics should include—

- The impact of PTE on unit and Soldier readiness.
- Common PTEs for units and Soldiers.
- An overview of TEM.
- Components of TEM.
- Normal responses to PTEs.
- Triggers to refer Soldiers for BH evaluation.
- Development of SOPs for responding to PTEs when they occur.

6-8. Preemptive consultation and education prepare unit leaders to institute TEM interventions following a PTE. The 5 R's are a good model to build on. In acute TEM interventions, leaders should consider interventions that target—

- Safety, security, and survival.
- Food, hydration, clothing, and shelter.
- Sleep.
- Medication (replace medications destroyed or lost).
- Orientation of unit/Soldiers to developing situation.
- Restoration of communication with unit, dependents, friends, and community.
- Location should be at a secure site that provides protection from ongoing threats, toxins, and harm.

6-9. Consultation and education to Soldiers should emphasize normalizing the common reactions to trauma, improving their coping skills, enhancing self-care, facilitating recognition of significant problems, and increasing knowledge of and access to COSC services. Post-PTE UNAs guide further consultation and education efforts. Combat and operational stress control providers should be aware that leaders may not have experienced the PTE in person, but have experienced the PTE through their Soldiers and may require support.

6-10. For an in-depth discussion of consultation and education, refer to Chapter 5.

Combat and Operational Stress Control Triage

6-11. The COSC provider should be prepared to provide COSC triage in the aftermath of the PTE. Consultation and education is important to ensure appropriate and timely referrals for triage. The COSC provider can offer the following guidelines in referral of Soldiers for COSC triage:

- Persistent or worsening traumatic stress reactions (such as dissociation, panic, autonomic arousal, and cognitive impairment).
- Significant functional impairments (such as role/work relationships).
- Dangerousness (suicidal or violent ideation, plan, and/or intent).

- Severe psychiatric comorbidity (such as psychotic spectrum disorder, substance use disorder, or abuse).
- Maladaptive coping strategies (such as pattern of impulsivity or social withdrawal under stress).
- New or evolving psychosocial stressors.
- Poor social support.
- Failure to respond to acute supportive interventions.
- Exacerbation of preexisting psychiatric conditions.
- Soldier request for assessment.

Stabilization

6-12. The COSC provider should be prepared to provide or coordinate stabilization services following the PTE. Pre-PTE coordination with medical unit personnel promotes safe management.

Soldier Restoration

6-13. The measures below are applicable to Soldiers with COSR following a PTE. The COSC provider should be familiar with the 5 R's and with BICEPS. In keeping with restorative efforts, the COSC provider focuses on the following measures through leadership consultation, Soldier education, and/or direct management:

- Minimizing exposure of Soldiers with COSR to further PTE.
- Reducing physiological arousal.
- Mobilizing support for those who are most distressed.
- Providing information and fostering communication and education.
- Using effective risk communication techniques.
- Proving assurance/reassurance.
- Mitigating fear and anxiety.
- Encouraging sleep hygiene.
- Reestablishing routines.
- Promoting exercise and nutrition.
- Encouraging self-paced emotional ventilation.
- Discouraging use of alcohol/substances.

Behavioral Health Treatment

6-14. Given the correlation between PTE and development of traumatic stress disorders the COSC provider must be familiar with the best current practices for evaluation and treatment. The Veterans Health Administration (VHA)/DOD Clinical Practice Guidelines website (http://www.oqp.med.va.gov/cpg/cpg.htm) offers clinicians evidence-based assessment and treatment algorithms for acute stress disorder, PTSD, and many other BH disorders.

6-15. In recent years, the use of early interventions in response to PTE has come under critical review. The focus of much of this debate is on the use of psychological debriefing (PD) and more specifically, the critical incident stress debriefings provided to individuals or groups exposed to PTEs. Current research suggests that PD/critical incident stress debriefings can be harmful to participants, while failing to reduce traumatic stress reactions or to prevent the progression to PTSD. As PD/critical incident stress debriefings undergo definitive study, COSC providers should adhere to current evidence-based best practices.

6-16. In accordance with the VHA/DOD Clinical Practice Guidelines (http://www.oqp.med.va.gov/cpg/cpg.htm) for PTSD, COSC providers should—

- Consider alternative methods to PDs for individuals affected by PTEs.
- Avoid PD as a means to reduce acute posttraumatic distress (acute stress reaction or acute stress disorder) or to slow progression to PTSD.

- Understand there is insufficient evidence to recommend for or against conducting structured group debriefings.
- Be aware that compulsory repetition of traumatic experiences in a group may be counter-productive.
- Consider group debriefings with preexisting groups (such as teams, units, emergency medical treatment [EMT] teams, coworkers, family members) may assist with group cohesion, morale, and other important variables that have not been demonstrated empirically.
- Emphasize that group participation must be voluntary.

LEADER-LED AFTER-ACTION DEBRIEFING

6-17. A leader-led after-action debriefing is lead by a platoon, squad, or team leader and is not normally conducted above platoon level. The leader-led after-action debriefing should be conducted after all missions especially when the maneuvers did not go according to plan. A leader-led after-action debriefing may even be sufficient for PTEs involving injury or death. For the leader to conduct a PD, his personnel should have received previous PDs; and normally provide peer support and validation for showing and talking about their emotional reactions during and after the debriefing. The best time to conduct this debriefing is as soon as is feasible after the team/squad/platoon has returned to a relatively safe place and members have replenished bodily needs and are no longer in a high state of arousal. Usually a well-conducted leader-led after-action debriefing is the best option to manage PTEs during a mission. The exception to this type of debriefing is when the event evoked reactions that seriously threaten unit cohesion and/or have a high likelihood of arousing disruptive behavior and emotions. In these situations the leader should ask himself the following—

- Should I conduct the debriefing?
- Should a trained facilitator be present?
- Should a request for COSC TEM be submitted for his team/squad/platoon?

CONDUCTING A LEADER-LED AFTER-ACTION DEBRIEFING

6-18. These debriefings require the leader to extend the lessons-learned orientation of the standard AAR. He uses the event reconstruction approach or has the individuals present their own roles and perceptions of the event, whichever best fits the situation and time available. See Training Circular (TC) 25-20 for definitive information on AARs. When individuals express or show emotions, the leader and the teammates recognize and normalize them; they agree to talk about them later and support the distressed Soldier through personal interactions. The group then returns to determining the facts. Lessons-learned discussion is deferred until all the facts are laid out. See FM 22-51 for additional information. The leader may provide education about controlling likely reactions or referral information at the end, depending on his knowledge and experience.

6-19. When a PTE is likely to create individual or collective guilt, distrust, or anger, the unit leader should be encouraged to request COSC assistance. Either a COSC or a UMT person trained in TEM sits in with the leader-led debriefing as a familiar and trusted friend of the unit. The COSC or UMT facilitator helps the unit/team leader rehearse and mentors the leaders on the debriefing process. During the PD, the facilitator can ask questions of the group to clarify the facts and steer the process away from divisive anger, blaming, and scapegoating toward a positive, cohesion-restoring outcome. This method is halfway between a simple leader-led after-action debriefing and a PD and is referred to as a *facilitated leader-led after-action debriefing*. The leader conducting the debriefing must be attentive to identify individuals needing COSC follow-up.

6-20. Leaders in positions above platoon level also have a role in leader-led after-action debriefing. Company commanders and 1SGs may conduct leader-led after-action debriefing with their subordinate leaders. Battalion commanders and higher may conduct leader-led after-action debriefings with their staff after distressing actions and may include subordinate leaders when time allows bringing them together.

COOL DOWN MEETINGS

6-21. A cool down meeting is referred to as an immediate, short meeting when a team or larger unit/group returns from the battlefield or other missions. These cool down meetings are held after heavy/intense battles with the enemy or a shift in the mission has occurred which is highly arousing and/or distressing. This is especially important after critical events. The cool down meeting is an informal event and occurs before the participants fully replenish their bodily needs and precedes any other activities including more COSC interventions, or return to the mission.

6-22. Personnel who coordinate and wait for the return of the unit that has been in a heavy intense battle, include leaders or supporting officers or NCOs from the command, UMT, and COSC providers. (In domestic support operations the same personnel identified above or other trained personnel from governmental or nongovernmental organizations such as the Red Cross may be waiting on the return of the unit after a PTE. These personnel may be present at the cool down meeting.)

COMPONENTS OF A COOL DOWN MEETING

6-23. Components of a cool down meeting may include—

- Assembling all of the unit personnel at a safe and relatively comfortable location for a brief period of time (about 15 minutes).
- Receiving or sharing nonstimulating beverages and convenient food (*comfort* foods if available).
- Providing personnel the opportunity to talk among themselves.
- Giving recognition and praise for the difficult mission they have completed.
- Providing information to unit personnel on where and how they will rest and replenish.
- Previewing the immediate agenda for the unit on what will happen after the cool down meeting including plans for further debriefing and/or other available stress control or morale/welfare intervention.
- Providing announcements pertaining to further preparations and expected time of return to the mission.

6-24. The COSC personnel may have a role in cool down meetings. In consultation and education, COSC consultants emphasize the value of cool down meetings and the simplicity of the components, which are easily neglected in crisis situations. When feasible, they unobtrusively attend the cool down meeting, showing "presence" while learning of the event, getting familiar with the key people, and observing anyone showing signs of distress and being available to them. If requested, they may give a very brief introduction and review of normal stress reactions that unit personnel may have in the next few hours.

Chapter 7

Combat and Operational Stress Control Support for Units Undergoing Reconstitution

UNIT RECONSTITUTION SUPPORT

7-1. Unit reconstitution support is defined in FM 100-9 as extraordinary actions that commanders plan and implement to restore units to a desired level of combat effectiveness commensurate with mission requirements and available resources. Besides normal support actions, reconstitution may include—

- Removing the unit from combat.
- Assisting the unit with external assets.
- Reestablishing the chain of command.
- Training the unit for future operations.
- Reestablishing unit cohesion.

7-2. When tasked as part of a reconstitution task force, COSC personnel are responsible for providing units with Soldier restoration, performing the COSC functions of the UNA, and providing consultation and education, as required. They provide triage, stabilization, Soldier restorations, and short time-constrained COSC treatment, when needed. Reconstitution is a time-constrained process, but Soldier restoration may be provided at the reconstitution site if several days are available. The focus of consultation and education efforts include—

- Rebuilding unit cohesion.
- Integrating new Soldiers or groups of Soldiers into the unit.
- Facilitating assumptions of command by replaced leaders.
- Facilitating the building of Soldiers' confidence in their leaders and themselves.
- Mentoring unit leaders in leader-led after-action debriefing process.
- Advising on COSC aspects of bereavement memorial services and communication with the family support group and unit families. The COSC personnel also facilitate leader-led after-action debriefings and lead or serve as observers in TEM interventions.

RECONSTITUTION PROCESS

7-3. Reconstitution of units transcends normal day-to-day force sustainment actions. Reconstitution is a total process. Its major elements are reorganization, assessment, and regeneration, in that order. All COSC personnel should be thoroughly familiar with FM 100-9.

Reorganization

7-4. Unit reorganization primarily involves a shifting of internal resources and is accomplished as either immediate or deliberate reorganization. The commander of the attrited unit decides to reorganize when required and may consider the following—

- Immediate reorganization is the quick and usually temporary restoring of degraded units to minimum levels of effectiveness. Normally, the commander implements immediate reorganization at his combat location or as close to that site as possible to meet near-term needs. The COSC personnel provide consultation with key POC in the unit (leaders, UMTs, and trauma/health care specialist, usually by telecommunication).
- Deliberate reorganization is done to restore a unit to the specified degree of combat effectiveness. Usually, more time and resources are available further to the rear. Procedures are similar to immediate reorganization except that some personnel and weapons system

replacement resources may be available, equipment repair is more intensive, and more extensive cross- leveling is possible. When used in reorganization, cross-leveling involves the movement of personnel and/or equipment between units to achieve equalization. The process is accomplished while maintaining or restoring the combat effectiveness of the units involved.

- The role of COSC personnel in deliberate reorganization may require a COSC team to deploy to the reorganization site. They assess the stressors and stress reactions and advise the commanders on supportive actions, (such as those for the reorganization of small unit-level elements), assists command with transitions and integration of new replacements.
- When the reorganization involves battalion size or larger units the process becomes more like the coordinated operation described for unit regeneration, provided below. It is likely to draw upon fitness team expertise and Soldier restoration assets from rearward as well as on the forward deployed teams.

REGENERATION

ASSESSMENT PHASES

7-5. The defining characteristic of regeneration is the massive infusion of personnel, equipment, and assistance under the directions of higher headquarters. Assessment measures a unit's capability to perform its mission. It occurs in two phases. The unit commander conducts the first phase. He continually assesses his unit before, during, and after operations. If he determines it is no longer mission-capable even after reorganization, he notifies his commander. The higher headquarters either changes the mission of the unit to match its degraded capability or removes it from combat. External elements may also have to assess the unit after it disengages. This is the second phase. These elements do a more thorough evaluation to determine regeneration needs. They also consider the resources available.

7-6. The second phase of assessment begins with an initial survey by a team sent by the higher headquarters. This team determines the status and needs of the attrited and exhausted unit as it moves to the regeneration site. Some of the key issues in estimating the COSC needs of the unit include the—

- Percentage and nature of casualties.
- Duration of operations and environmental exposure.
- Status of hydration, nutrition, and sleep.
- Loss and current effectiveness of leaders.
- Attitudes, perceptions, and level of confidence of unit survivors.
- Traumatic events.

7-7. The COSC consultants at every level must provide consultation to all command surgeons on the importance of including BH/COSC personnel in all reconstitution operations. The initial evaluation team should include a subordinate commander COSC consultant. Task-organized CSC teams normally deploy to the unit during the second phase to provide UNA, consultation, and other COSC activities and interventions. Unit needs assessment with high formality may be requested by the higher headquarters for decision-making purposes. These assessments are feasible for selected units, with prior coordination, as much of the logistical complexity is reduced by the orchestrated planning at higher command echelons.

Conducting Regeneration

7-8. Regeneration involves rebuilding a unit through large-scale replacement of personnel, equipment, and supplies. If required, it includes internally reorganizing; reestablishing or replacing essential C2 and conducting mission-essential training for the reconstituted unit. Regeneration is required when heavy losses of personnel and equipment leave a unit combat-ineffective and unable to continue its mission. The commander controlling assets to conduct regeneration decides whether to use resources for this purpose. Regeneration has two variations: incremental regeneration or whole-unit regeneration. Incremental regeneration is the massive infusion of individual personnel replacements and single items of equipment into the surviving unit elements. Whole-unit regeneration is the replacement of whole units or definable subelements, such as squads, crews, and teams. The S1/G1, S4/G4, and medical staffs coordinate the dispatch of the regeneration task-force teams. These teams occupy the reconstitution site before arrival of the exhausted unit. The COSC assets that perform the indicated initial assessment are needed as part of this task force. The reconstitution task force guides each element of the arriving units into its designated areas. The regeneration task force provides for the immediate needs of the survivors. This should include personal gear and equipment to replace lost or damaged items, food services, personnel services, maintenance teams, and medical teams to provide sick call services while organic medical personnel rest. Replacement personnel are sent to the reconstitution site. The COSC personnel assist with their assimilation into the regenerated unit.

RECONSTITUTION RESOURCE REQUIREMENTS FOR COMBAT AND OPERATIONAL STRESS CONTROL

7-9. Factors which influence the resource requirements for COSC in reconstitution (deliberate reorganization and regeneration) include the size of the unit, number of subunits which have suffered heavy casualties, the extent of emotional trauma, and time available. The focus of COSC for reconstitution support was provided in Paragraph 7-2. A guideline is provided in Table 7-1 for COSC personnel requirements. This staffing should not overly rely on organic COSC staff that will be in need of rest.

Table 7-1. Reconstitution operations guideline for combat and operational stress control personnel requirements

Size of Unit	*Personnel Required*
Company/Troop	2 to 4
Battalion/Squadron	6 to 12
BCT/Brigade Regiment	12 to 30
Division	30 to 60

Chapter 8

Combat and Operational Stress Control Triage

SECTION I — THE TRIAGE PROCESS FOR COMBAT AND OPERATIONAL STRESS CONTROL

SOLDIER TRIAGE

8-1. The COSC triage process is the sorting of Soldiers based on an assessment of their needs and capabilities, and the location where they can best be managed in keeping with BICEPS principles. Triage is applicable at every level of care. The two key components of COSC triage are assessment and disposition.

8-2. Assessment is an evaluation of the Soldier's physical and BH needs, potential medical emergencies, and other safety risks. Assessment is performed by COSC personnel according to their professional training, expertise, and standards.

8-3. Disposition is the COSC intervention plan to address the needs identified in the assessment. Disposition has two components that include—

- Determining what intervention techniques best address the Soldier's needs and functional capabilities.
- Selecting the best location where the Soldier can be managed. The personnel conducting COSC triage should consider the needs, abilities, and the safety of the Soldier. They should also consider the unit's capacity to provide COSC interventions based on its OPTEMPO mission, resources, response to prior consultations, and willingness to participate in COSC interventions.

TRIAGE ALGORITHM FOR COMBAT AND OPERATIONAL STRESS CONTROL

8-4. Like surgical triage categories, COSR also uses triage categories. Each of the COSR triage categories are discussed in detail in the below paragraphs. The COSC triage algorithm presented in Table 8-1 uses some of the triage categories.

Table 8-1. The combat and operational stress control triage algorithm

<table>
<tr><td>Step 1</td><td colspan="5">Is this a medical emergency?</td></tr>
<tr><td></td><td>Yes</td><td>Refer to nearest MTF</td><td>No</td><td colspan="2">Go to Step 2</td></tr>
<tr><td>Step 2</td><td colspan="5">Does the Soldier require medical/behavioral observation?</td></tr>
<tr><td rowspan="3"></td><td rowspan="3">Yes</td><td rowspan="3">Go to Step 3 A</td><td rowspan="3">No</td><td colspan="2">Does Soldier have presumptive COSR or MH disorder?</td></tr>
<tr><td>Yes</td><td>Go to Step 3 B</td></tr>
<tr><td>No</td><td>Help-in-place</td></tr>
<tr><td>Step 3</td><td colspan="2">A. Can the Level II MTF or CSC Soldier Restoration Center provide adequate evaluation and intervention?</td><td colspan="3">B. Can the Soldier's unit support the 5 R's or other treatment interventions?</td></tr>
<tr><td rowspan="4"></td><td>Yes</td><td>HOLD*</td><td>Yes</td><td colspan="2">UNIT</td></tr>
<tr><td rowspan="3">No</td><td rowspan="3">REFER</td><td rowspan="3">No</td><td colspan="2">Is there a suitable support unit?</td></tr>
<tr><td>Yes</td><td>REST</td></tr>
<tr><td>No</td><td>HOLD*</td></tr>
<tr><td colspan="6">* When deciding between two or more potential Level II MTFs or CSC Restoration Centers, refer the Soldier to the one closest to his unit that meets his COSC needs.</td></tr>
<tr><td>Step 4</td><td colspan="5">Use on subsequent triages. Has the Soldier improved after appropriate duration of intervention?</td></tr>
</table>

TRIAGE CATEGORIES FOR COMBAT AND OPERATIONAL STRESS REACTION CASES

8-5. The following are triage categories that may be used for COSR cases. *Help-in-place* (HIP), *rest*, *hold*, and *refer* cases are discussed below.

HELP-IN-PLACE CASES

8-6. *Help-in-place* is used to identify those cases that do not have severe COSR or BH disorders. They are provided COSC consultation and education, as appropriate, and remain on duty. These interactions may occur in any setting (for example, dining facility, workplace, or the post exchange). Individual identifying information is not retained or documented. There is no implicit or explicit therapist-patient or therapist-client relationship in HIP interactions.

8-7. The unit identifies those cases that remain with or return to their original unit, either for full duty with their section/platoon or for light duty with extra rest and replenishment within a headquarters element. This option depends on the unit's mission, resources, and the Soldiers symptoms. Personnel performing triage must, therefore, be familiar with the unit's situation and take that into account. When the Soldier's condition improves, the Soldier and/or unit may not feel that additional triage is necessary.

Note. *Help-in-place* is a new triage category that replaces the old *duty* category and is also used for unit cases.

REST CASES

8-8. *Rest* identifies those cases that are provided rest and replenishment in a nonmedical support unit, usually one that is in support of their unit. These Soldiers do not require close medical or BH observation or treatment. They are unable to return immediately to their own unit either because their unit cannot provide an adequate environment for the 5 R's; or transportation is not available for at least a day; or the 5 R's can best be coordinated from the nonmedical support unit. This option depends on the resources and mission of the available CS/CSS units, as well as on the Soldier's symptoms. Someone must be designated to be in charge of the Soldier and ensure that the 5 R's are provided. There must be a reliable transportation link to return the Soldier to his original unit after a day or two. When the Soldier's condition improves sufficiently for him to return to his unit, the Soldier and/or the supporting unit may feel that additional triage is necessary.

HOLD CASES

8-9. *Hold* refers to those cases that require close medical/BH observation and evaluation at a Level II MTF or COSC Soldier restoration center because the Soldiers symptoms are potentially too disruptive or burdensome for any available CSS unit or element. Soldier's symptoms may be caused by a BH disorder that could suddenly turn worse and require emergency treatment. The Level II MTF or COSC Soldier restoration center must have the capability to provide the necessary medical observation, diagnostic tools, and adequate stabilization for emergency treatment. When deciding among capable Level II MTFs or COSC Soldier restoration centers, refer the Soldier to the one closest to his unit that meets his COSC needs. Assessment of closeness considers speed and reliability of transportation and back. Consider transferring to another Level II MTF or COSC Soldier restoration center with increased capabilities before changing a Soldier's triage category to *refer*. All *hold* cases will be triaged again by COSC personnel or other trained medical personnel after they have been placed in this category.

REFER CASES

8-10. *Refer* cases are similar to the *hold* cases, except that *refer* cases are too disruptive and burdensome for the MTF or the COSC Soldier restoration center that is not resourced to care for this particular case. The MTF or COSC Soldier restoration center cannot provide the necessary level of diagnostic and treatment capabilities. *Refer* cases requiring care at a COSC reconditioning center, a Level III MTF or higher levels of care will be triaged by COSC or other trained medical personnel prior to being transferred to these facilities.

TRIAGE PERSONNEL FOR COMBAT AND OPERATIONAL STRESS CONTROL

8-11. All COSC personnel participate in the triage process according to their professional training, experience, and standards. Familiarization training among BH disciplines extends the effectiveness of all COSC personnel in triage skills. Medical care providers must be mentored to use the COSC triage process. A commander providing the 5 R's and placing a Soldier in a support unit for a temporary break does not equate to COSC triage.

TRIAGE CONSIDERATIONS

8-12. Triage should be initiated when the—

- Soldier is a self-referral.
- Chaplain has referred the Soldier.
- Medical personnel have requested a COSC consultation and referred the Soldier.

- Unit member/buddy has referred the Soldier.
- Leader has requested an informal referral.
- Soldier is a command-directed referral (see DODD 6490.1 and DOD Instruction [DODI] 6490.4).
- Combat operational stress control personnel observe a Soldier's behaviors which indicate possible COSR or a BH disorder.

8-13. Factors that influence an assessment may vary in depth and duration due to several other factors. These factors may include the—

- Referral source.
- Nature of the complaint.
- Observed needs.
- Medical/psychiatric history.
- Availability of resources.
- Amount of privacy for conducting assessment.
- Operational Tempo.
- Environmental conditions.
- Professional training of the person making the assessment.
- Command interest.
- Soldier cooperation.

Note. Regardless of these factors, COSC personnel are responsible for conducting the assessment in a timely manner within professional standards of practice.

DOCUMENTATION

8-14. An assessment is documented according to AR 40-66 whenever the Soldier—

- Is diagnosed with a BH disorder.
- Has a condition (or suspected condition) requiring emergency medical evaluation or treatment.
- Is prescribed medication.
- Is assessed to be a high safety risk (for example, homicidal or suicide, cognitive impairment, substance abuse, and impulsivity).
- Requests that documentation of his assessment be made in his medical records.
- Is evacuated beyond Level II MTF for further assessment or treatment.
- Is command-referred for a BH evaluation.

TRANSFER AND EVACUATION

8-15. All COSC personnel are responsible for knowing the transfer/evacuation policies and procedures within their AO. Policies and procedural information are available through the command surgeon, medical regulating officer (MRO) of the medical command, control, communications, computers, and intelligence (C4I) headquarters or major MTF. All relevant background and/or clinical documentation must accompany the Soldier during the transportation or evacuation process.

MODE OF TRANSPORTATION

8-16. Nonambulance transport is the preferred mode of transportation for COSR and nonurgent BH cases. Examples of nonambulance transportation include the Soldier's unit vehicles, supporting supply/logistics vehicles, and nonambulance medical vehicles. Ambulances convey patient status on Soldiers and often must be reserved for medical emergency. Under the provisions of the Geneva Conventions, ambulances must be used exclusively in the performance of humanitarian duties, therefore, they cannot be used to return Soldiers to duty and transport must be provided by the supported unit.

ESCORT

8-17. Depending on a Soldier's condition, an escort (either medical or nonmedical) may be necessary to provide safety, monitoring, and accountability during transportation or evacuation. The escort should be an NCO or officer of equal or greater rank/grade as that of the escorted Soldier. Escorts must be emotionally mature, responsible, and capable of conducting their escort duties. Frequently escorts carry the Soldier's clinical documentation to the destination MTF.

FEEDBACK

8-18. Good communication is essential for effective continuity of care during the transportation/evacuation process. The Soldier's unit must be informed about his location and status throughout the process. The sending party must provide sufficient documentation about the Soldier's condition, history, and administered interventions. The receiving party must provide feedback to the sending party regarding receipt of the Soldier and his documentation.

SECTION II — PRECAUTIONS AND DIFFERENTIAL DIAGNOSTIC PROBLEMS ASSOCIATED WITH COMBAT AND OPERATIONAL STRESS CONTROL TRIAGE

PRECAUTIONS FOR COMBAT AND OPERATIONAL STRESS CONTROL TRIAGE

8-19. Medical emergencies must be identified during COSC triage. Medical emergencies consist of physical illnesses or injuries and/or BH disorders that can result in permanent injury, disability, or death. Early identification of a medical emergency avoids unnecessary delay in definitive treatment. Medical emergencies can cause emotional and/or BH changes and may resemble COSR in presentation. The following conditions and behaviors could be medical emergencies. Therefore, it is important to ensure medical examinations and disposition of Soldiers that display the following conditions—

- Psychosis.
- Mania.
- Alcohol withdrawal.
- Substance intoxication.
- Delirium.
- Suicidal gesture, attempt, or high risk for suicidal behavior.
- Catatonia.
- Significant paresis, paralysis, and/or sensory loss.

8-20. Assessing for physical illnesses or injuries is a critical part of COSC triage. The COSC personnel must always consider physical illnesses or injuries that resemble COSR or BH disorders. Physical illnesses or injuries may not reach the threshold of a medical emergency, but must be recognized and appropriately treated. Assessing for physical illnesses or injuries requires an adequate review of body systems and a quick physical examination. The examination includes vital signs, examination of head, eyes, ears, nose, throat, chest, abdomen, and extremities with simple testing of reflexes and muscle strength. In field situations (Level I and II), negative or normal findings need to be documented on DD Form 1380 (US Field Medical Card). Refer to AR 40-66 for detailed information on patient accountability and management of individual health records. Any positive findings from the physical examination must be evaluated further. If the examiner has not checked various body systems, it is not reassuring to tell a Soldier that his physical or BH complaints are only COSR. All COSC personnel should receive familiarization training on basic medical examination techniques and in documenting medical information. Whenever a physical illness or injury is suspected, personnel should consult with their medical peers for further assistance. Some cases will require direct medical examination by a physician or physician assistant (PA). The COSC personnel should not order tests or procedures that do not directly influence case management. Medical tests may promote the patient role in the mind of the Soldier. Needless tests may delay a Soldier's RTD and encourage secondary gains.

DEFER DIAGNOSIS OF BEHAVIORAL DISORDERS

8-21. During assessment, COSC personnel must always consider BH disorders that resemble COSR, but defer making the diagnosis. The COSC personnel favor this default position to preserve the Soldier's expectations of normalcy (according to BICEPS). This is also done to avoid stigma associated with BH disorders and to prevent the Soldier identifying with a patient or sick role. Deferral is also preferred because some diagnoses require extensive history collection or documentation that is unavailable during deployment situations (such as personality disorders and attention deficit hyperactive disorder). It is possible that a Soldier can have a combination of COSR, BH disorders and physical illnesses/injuries at the same time. In such cases, COSC personnel must rely on their clinical experiences, training, and consultation with peers and medical personnel to distinguish among these sometimes overlapping conditions. Physical injuries/illnesses are treated at a MTF, however, the Soldier may return for further COSC interventions and activities. Deferral of diagnosis is preferred, but diagnosis can be considered if the Soldier—

- Presents for reemerging symptoms of a previously diagnosed and/or treated BH disorder.
- Presents for refill prescription of psychotropic medication.
- Has a medical condition or BH disorder listed above in paragraph 8-19.
- Is enrolled in a reconditioning program.
- Fails to improve after having received four to five days of continuous COSC interventions and activities at hold or refer status.
- Requires individual BH treatment.
- Is referred for multiple episodes of COSR.

DIAGNOSTIC CONSIDERATIONS FOR DIFFERENTIAL DIAGNOSTIC DISORDERS

Low-Grade Environmental or Stress-Related Illnesses

8-22. Low-grade environmental or stress-related DNBI illnesses can drain the Soldier's strength and confidence. For example, chronic diarrhea and slight fever may exhaust, demoralize, and contribute to COSR among Soldiers. These conditions should be treated medically, concurrently with physical replenishment, rest, reassurance and organized activities, which restore the Soldier's confidence. If they persist in spite of rest and symptomatic treatment, a more aggressive workup and treatment may be indicated.

Dehydration

8-23. Dehydration deserves special mention because it can be very insidious. Soldiers under battlefield or heavy work conditions become extremely dehydrated without feeling thirsty. This is especially likely in CBRN equipment, or in a desert/arctic environment. Refer to FM 4-25.10 for additional information on the prevention of dehydration and FM 4-25.11 for first aid measures.

Hyperthermia

8-24. Hyperthermia (overheating) in an otherwise healthy individual often first causes mild elation and excessive energy. This may be followed by irritability, disorientation, and confusion. When core body temperature climb above 106° Fahrenheit (F) or 41° Centigrade (C), the Soldier may become belligerent, combative, and have visual hallucinations. If brain temperature rises further, the Soldier collapses and convulses in heatstroke. Refer to FM 4-25.11 for first aid measures and FM 4-25.10 for additional information on prevention and first aid measures for heat injuries.

Hypothermia

8-25. Hypothermia may cause an individual to become disoriented when core body temperature falls below 95°F (34.6°C). The person may move and speak slowly. His skin looks and feels warm, leading him to take off clothing. He becomes disoriented, then unresponsive and may appear to be dead. Hypothermia is as likely in cool or even warm, wet climates as it is in extremely cold ones. Refer to FM 4-25-10 for additional information on prevention and FM 4-25.11 for first aid measures.

Overuse Syndromes

8-26. Overuse of muscles, joints, and bones that have not been prepared for the strain of field duties can lead to persisting stiffness, pain, swelling, and orthopedic injuries. If severe, these injuries may require evacuation to a hospital for evaluation. Even if these injuries are avoided, the unfit person who overexerts has days of stiffness, aching, and weakness. Such cases are likely to develop COSR if further demands are made on them.

Rhabdomyolysis

8-27. Rhabdomyolysis is one potentially dangerous complication of severe muscle overuse (and of heatstroke or crush injuries) in which myoglobin from damaged muscle cells injures the kidneys. It can cause fatigue, seizures, muscle tenderness, and muscle aches. A warning sign is dark (tea-colored) urine, but without laboratory testing, this is not easily distinguished from the concentrated urine of dehydration. Rhabdomyolysis is a medical emergency.

Head Trauma

8-28. Concussion may stun the individual and cause amnesia, residual confusion, and/or impulsive behavior. For any case of suspected head trauma or for any case of significant memory loss (especially for a discrete period of time), check scalp, eyes, ears, nose, cranial nerve signs, and vital signs for evidence of head injury. Cases with deteriorating mental status are medical emergencies. If one pupil becomes larger than the other, it is an extreme emergency requiring immediate hospitalization. Left untreated, the condition can progress rapidly to coma and respiratory arrest within hours. If a head injury is suspected, monitor mental status and vital signs periodically, especially respiration, even though physical findings are negative. Awaken the Soldier periodically to check mental status and pupil size (allowing sufficient time to recover from any grogginess on awakening). Continuous monitoring is appropriate if there are serious concerns about the risk.

Spinal Cord Trauma

8-29. Pressure, bruising, and hematomas of the spinal cord, as well as severing of the spinal cord, can cause spinal shock, with loss of sensory and/or motor functions below the level of the injury in the affected dermatome and muscle group patterns. The loss of function may be bilateral, unilateral, or partial. These cases could be confused with paralysis or sensory loss presentations of COSR. Further manipulation of a fractured spine can worsen or make permanent the spinal cord damage. Information from the history of onset, a cautious physical and neurological examination, or complete relief of symptoms following hypnosis or strong positive suggestions could demonstrate convincingly that this is only COSR. It is best to be cautious and keep the spine immobile during care and transportation.

Postconcussion Syndromes

8-30. Postconcussion syndromes may persist weeks to months beyond the period of acute concussion. Postconcussion syndromes may include perceptual or cognitive impairment, poor impulse control, and difficulty in planning ahead. These are often accompanied by cranial nerve deficits or soft neurological signs.

Abdominal Trauma

8-31. Ruptured spleen or other intraperitoneal bleeding may cause shock. The Soldier may arrive in a fetal position and be unresponsive but have reflex guarding due to peritonitis.

Air Emboli and Focal Brain Ischemia

8-32. High blast overpressures from incoming high explosive ordnance can produce air emboli (bubbles in the blood) and focal brain ischemia (small areas in the brain which cannot get oxygen because the blood flow has been interrupted). Nuclear explosions can do this, as can high explosives when shock waves are magnified by reflection within bunkers, buildings, and trenches. Ruptured eardrums, general trauma, and evidence of pulmonary damage should be detectable. Cases may have stroke symptoms (loss of muscle strength, loss of sensation in parts of the body, and/or speech disturbances), which may be subtle and mistaken for COSR.

Laser Eye Injury

8-33. Laser range finders/target designators cause small burns on the retina if they shine directly into the eye, even at great distances and especially if viewed through optics. If the laser beam causes a small retinal blood vessel to bleed inside the eyeball, the person will see red. If blood inside the eye is confirmed on examination, the Soldier should be evacuated to a hospital with verbal reassurance that he may RTD soon. If the laser does not hit a blood vessel, the Soldier may see only flashes of light, followed quickly by some painless loss of vision. If the laser damage areas of the eye responsible for peripheral vision, the Soldier may never recognize a visual deficit. If the Soldier was looking directly at the laser source however, there will likely be a major loss of visual clarity. With simple retinal burns in the retina's periphery, most of the visual symptoms recover with hours to days of rest, reassurance, and nonspecific treatment the same as with COSR. Calm, professional treatment at each level of medical care should emphasize that the injury is not vision-threatening and the chances for some, if not total, recovery is good. Soldiers with the simple retinal burns should provide self-care to decrease the risk of assuming a patient mind-set and to promote their chances of RTD. For additional information on the management of laser eye injuries, refer to FM 8-50.

Middle Ear Injuries/Diseases

8-34. Temporary loss of hearing can be cause by a decreased acoustic sensitivity following a brief extremely intense noise (explosive) or less intense, longer duration noise. Tinnitus (ringing in the ears) can also result from acoustic nerve damage or irritation, as well as from high doses of certain drugs, such as aspirin. Distinguishing physiologic from psychogenic hearing loss may require consultation with an otolaryngologist (ears, nose, and throat specialist).

Peripheral Neuropathies

8-35. Peripheral neuropathies include compression neuropathies, which are especially likely in military settings (for example, *rucksack palsy*). Depending on severity, they may require temporary job reclassification during convalescence. As these injuries are not life-threatening, a hasty diagnosis should not precede a trial of Soldier restoration treatment.

Uncommon Endemic Neurologic Disorders

8-36. These physical diseases can first manifest with cognitive emotional and/or behavioral symptoms. A comprehensive neurological examination is required for the definitive diagnoses. Examples include—

- Guillain-Barre Syndrome manifests with muscle paralysis, usually without sensory loss, which ascends the legs and arms, then the trunk, over hours to days. It is sometimes triggered by immunizations, as might be given to troops deploying overseas. It often progresses to a life-threatening situation as the muscles of respiration become involved. This requires evacuation to EAC and continental United States (CONUS). Fortunately, recovery is usually complete, but it takes months to years.

- Multiple sclerosis is a disease that can mimic many types of COSR/BH disorders with its sometimes transitory, shifting motor, sensory, speech, and cognitive/emotional symptoms. It is made worse by stress and may be difficult to diagnose. Once confirmed, multiple sclerosis cases should be evacuated to CONUS, as should other rare, progressive diseases like Lou Gehrig's disease (amyotrophic lateral sclerosis).
- True convulsive seizure can manifest after head injury or a sublethal or chronic nerve agent exposure. These are treated with normal anticonvulsant medications. Fear of nerve agent exposure may lead some Soldiers to experience psychogenic seizures. These psychogenic seizures are also called pseudo-seizures. In addition to falling unconscious and convulsing, urinary and fecal incontinence can occur during a pseudo-seizure.

SUBSTANCE ABUSE/DEPENDENCE

ALCOHOL

8-37. Substance abuse is an example of misconduct stress behaviors and not necessarily COSR. Drug and alcohol abuse may occur in active CZs and nearby areas where use is explicitly prohibited and severely punished. Personnel performing the COSC assessment should be familiar with evaluation and treatment of substances abuse and dependency. Combat operational stress control providers should consider the following:

- Heavy habitual use of alcohol, even by otherwise capable officers and NCOs, may go unnoticed in peacetime. However, alcohol abuse it may degrade necessary mission performance demanded by combat and may result in withdrawal symptoms when access to alcohol is interrupted. Alcohol withdrawal is potential medical emergency; consultation with medical personnel is essential.
- Intoxication or withdrawal from alcohol, barbiturates, and tranquilizers may be mistaken for COSR or another BH disorder. Intoxication or withdrawal requires medical treatment. Withdrawal seizures or impending or ongoing delirium tremens need emergency medical treatment.

OVERUSE OF STIMULANTS

8-38. Stimulants may cause panic attacks, hyperactivity, mania, rage attacks, psychosis or paranoia. Cessation of amphetamines after prolonged use causes a crash characterized by extreme sleepiness, lethargy, overeating depression and suicidal thinking. This condition may require one to two weeks of hospitalization to assure safe recovery.

HALLUCINOGENIC DRUGS

8-39. Hallucinogenic drugs cause sensory distortion, panic, bizarre thoughts, and potentially dangerous behaviors. These drugs may be employed by the enemy as chemical or biological warfare agents. Phencyclidine hydrochloride (PCP) is especially problematic since it also blocks pain and tends to make those under its influence paranoid, violent, and abnormally strong. Hallucinogenic drug psychosis should not be treated with antipsychotic drugs. Physically restrain and sedate patients as necessary.

INHALATION OF FUMES

8-40. Inhalation of fumes (either by accident or as deliberate abuse) and carbon monoxide poisoning can cause disoriented, abnormal behavior. Supportive treatment and, specific antidotes/medication may be needed.

ANTICHOLINERGIC DELIRIUM

8-41. In combat, atropine delirium may occur. Soldiers are equipped with atropine injectors to use as first aid against nerve agents. Two milligrams (mg) (one atropine injector) without nerve agent challenge can cause rapid pulse, dry mouth, slightly dilated pupils, decreased sweating (hot, dry, flushed skin), and urinary retention. In some individuals, 6 mg of atropine (equal to three atropine injectors) may cause hallucination and disorientation in the absence of a nerve agent challenge. Such side effects may be more pronounced in sleep-deprived Soldiers. Overheated Soldiers are more susceptible to the atropine side effects. Atropine compounds the complications of overheating by diminishing the body's ability to lose heat through sweating. One dose (2 mg) of atropine can reduce the efficiency of heat-stressed Soldiers. Two doses (4 mg) will sharply reduce combat efficiency, and 6 mg will incapacitate troops for several hours. Some plants can also cause anticholinergic delirium when eaten.

ANTICHOLINESTERASES

8-42. A nerve agent is an anticholinesterase similar to many insecticides. Low-dose nerve agent exposure may produce miosis (pinpoint pupils) without other signs. Miosis decreases vision except in very bright light and may cause eye pain when attempting to focus. This miosis may take hours to days to improve spontaneously, depending on the degree and type of exposure. Evidence gathered from farm workers poisoned by insecticides suggests that mild personality changes, insomnia, nightmares and chronic persistent depressive symptoms may be seen even after use of an antidote. Low-dose nerve agent exposure may lower the seizure threshold of many Soldiers. True epileptic seizure cases must be distinguished from those Soldiers who may have pseudo-seizures.

BEHAVIORAL DISORDERED PATIENTS IN THE THEATER

PRIMARY BEHAVIORAL DISORDERS

8-43. Primary BH disorders (especially schizophrenic-form/schizophrenic disorder, major depression, and bipolar disorder) will continue to occur at approximately the same rate as in peacetime. Some Soldiers may hide their disorders by receiving care through civilian channels. Once in the theater they may experience a relapse or self-refer to a MTF when their medication supply is exhausted.

PERSONALITY DISORDERS

8-44. Preexisting personality disorders may make a Soldier unable to adapt to military life. However, studies have failed to show a relationship between personality disorders and the likelihood of breakdown in combat. Once Soldiers with personality disorders have developed COSR or a BH disorder, they may have greater difficulty recovering and RTD. Diagnosis should never be made in haste; diagnostic criteria must be supported with adequate historical information.

Chapter 9

Combat and Operational Stress Control Stabilization

SECTION I — EMERGENCY STABILIZATION

STABILIZATION

9-1. Emergency stabilization is the acute management of disruptive behavior resulting from COSR and/or a behavioral disorder. The disruptive behavior severely impacts unit functioning by posing a danger to self and/or others. In some cases, an underlying medical condition leads to the disruptive behavior and may present an additional threat to the Soldier's life. Emergency stabilization consists of interventions that temporarily reduce a disturbed Soldier's threat of self or others, thereby allowing further medical evaluation and/or treatment. Some behavioral disorders are associated with violent behavior, such as psychotic disorders, bipolar manic disorders, antisocial personality disorder, and borderline personality disorder. Violent behavior is also associated with disruption of brain functioning due to organic factors such as intoxication, hyperthermia, metabolic imbalance, or CBRN exposure.

9-2. The COSC triage process will be repeated throughout the emergency stabilization and will determine the disposition of the Soldier. After emergency stabilization, subsequent triage could result in an immediate RTD, transfer to a COSC Soldier restoration program for observation, or further evacuation.

METHODS USED FOR EMERGENCY STABILIZATION

9-3. Methods that may be used for emergency stabilization include—

- Providing verbal reassurance and reorientation are the best methods for controlling an agitated or disruptive Soldier. If these fail, a nonthreatening show of strength or force may suffice, or sedating medications may be offered to the Soldiers. If all other means fail to reduce the threat to self and/or others, physical restraint must be considered. Given the risk for violence, it is inadvisable to attempt subdue/restraint method one-on-one.
- Applying physical restraints is reserved for subduing and restraining agitated or disruptive Soldiers who fail to respond to safer and less restrictive forms of restraint (for example, verbal warnings or show of strength). Placing a disturbed Soldier into physical restraint increases the risk of injury to the Soldier and restraint team. Prolonged or improper application of physical restraint can cause injury to the disturbed Soldier. Given the potential for injury, it is paramount that COSC personnel receive training in proper physical restraining methods. Safe medical evacuation by ground (preferred) or air ambulance.
- Regardless of the method, the restrained Soldier must be checked frequently to guard against nerve injuries or impaired circulation, which may lead to skin ulcers or gangrene. It is important to check periodically to ensure the Soldier is not secretly escaping from restraints. The Soldier is provided verbal reassurances with positive expectations for his recovery each time he is checked.
- Chemical restraints (such as medication) can be administered to a disturbed Soldier to reduce the risk of harm to self or others. Medication may be offered to the Soldier in conjunction with verbal reassurances and reorientation. Chemical restraints may only be ordered by a medical professional who is authorized to prescribe medication when a Soldier is incompetent to make medical decisions for himself and/or when the Soldier's behavior places himself or others in danger. Once administered, medical personnel must observe for side effects and adverse reactions, and must consider administering additional medications as needed. When a Soldier is in physical restraints, medication may no longer be essential, but may serve to reduce the risk of escape, limb damage, and overheating. As a secondary benefit, once the medications reduce the

Soldier's agitation, others in the vicinity may feel safer and calmer. Before prescribing an antipsychotic medication, there are a few things to consider. First, some antipsychotic drugs may take several hours or days to take effect. Second, early administration of an antipsychotic drug may confuse the clinical picture for the next evaluator in the evacuation chain. The recommendation for most cases is to use no medication unless it is truly necessary for management.

SECTION II — COMBAT AND OPERATIONAL STRESS CONTROL FULL STABILIZATION

FULL STABILIZATION

9-4. Full stabilization is normally the mission of the Level III MTF such as a CSH specialty clinic's psychiatric service. Full stabilization goes beyond securing the safety of the Soldier and those around him. It provides a safe environment for the Soldier to receive treatment interventions, continued evaluation, and assessment for RTD potential. If RTD within the evacuation policy is not feasible, the full stabilization process helps to prepare the patient for a safe, long-distance evacuation. If a Level III MTF is not available, full stabilization may be accomplished by COSC personnel when appropriately supported or by using equipment diverted from the COSC Soldier restoration capability. Considerations for full stabilization may include—

- Conducting full stabilization for NP patients is desirable for the sake of the Soldier's future treatment and for the potential of returning some Soldiers to duty. However, full stabilization is personnel intensive with a relatively low RTD payoff. Providing only sufficient stabilization to allow evacuation from the theater may be acceptable in order to maintain the other COSC functions.
- Ensuring appropriate timely evacuation of Soldiers with NP/behavioral disorders according to the theater evacuation policy. It is preferred that full stabilization is achieved for all NP patients to facilitate appropriate and timely evacuation according to the theater evacuation policy.
- Assessing and triaging of COSR Soldiers undergoing full stabilization is an ongoing process. In subsequent triages, if a Soldier becomes stable the potential for RTD, they may be transferred to a COSC Soldier restoration or reconditioning program, or may be RTD directly.

TENETS OF FULL STABILIZATION

9-5. The COSC full stabilization includes ongoing evaluation of RTD potential. This requires assessment of mental status and performance capability overtime without excessive drug effects or limitation on activity. Contact with the Soldier's unit may be needed to get information on prior history and functioning. The further from the unit the Soldier has been evacuated, the more difficult it may be to contact the Soldier's unit. Full COSC stabilization normally takes several days. To the extent compatible with safety, the stabilization program should adhere to the principles and methods for treating COSR and behavioral disorder (such as BICEPS and 5 R's).

9-6. During full stabilization, special efforts should be made to maintain and reinforce the Soldier's identity as a Soldier. Techniques that may be helpful in maintaining the Soldier mind-set include—

- Keep Soldiers in duty uniform, not pajamas, as soon as they can be allowed.
- Maintain rank distinctions and appropriate military courtesy.
- Encourage self-care and helping behaviors.
- Engage in military work activities appropriate to the Soldiers' level of function and MOS.

9-7. Initial and ongoing assessments are essential to tailor the treatment to the Soldier's individual needs. It is essential that clinical documentation is available for these assessments. The Soldier's condition is an evolving one, and must be monitored throughout full stabilization. If assessed capable to RTD, efforts should be made to return the Soldier to duty.

9-8. Ongoing treatment and/or therapeutic modalities are essential to improving a Soldier's chances to RTD whether in theater or after evacuation. Therapeutic modalities are similar to those used on inpatient units, but must remain consistent with COSC principles. These modalities include medication, individual psychotherapy, group psychotherapy, and appropriate therapeutic occupations. Observed responses to therapeutic modalities allow informed recommendation for RTD status.

FULL STABILIZATION FACILITIES

9-9. Full stabilization is commonly conducted in the CSH. The CSH can provide more sophisticated procedures, laboratory and x-ray capabilities than are available at a medical company MTF. If a medical unit, CSC is providing emergency or full stabilization, the Soldier must be kept separate from other Soldiers in Soldier restoration or reconditioning.

9-10. If the Level III MTF (CSH) cannot provide sufficient inpatient psychiatric treatment for Soldiers requiring stabilization and preparation for evacuations, the theater/AO COSC consultant may recommend up to two temporary COA until the shortfalls resolve. He can recommend to the higher medical command that COSC personnel from one or more medical unit, CSC augment the Level III MTF psychiatric service until the caseload decrease or until replacements or additional COSC personnel are brought into theater or the MTF. Lastly, the medical C2 headquarters may direct that a COSC Soldier restoration or reconditioning assets be collocated with the Level III MTF to provide an "overflow" ward as well as augmenting staff. Definitive information is provided on the CSH NP ward staff capabilities in FM 4-02.10.

9-11. The MF2K CSH has a psychiatrist, three psychiatric nurses and a medical/surgical nurse, a social worker, nine mental health specialists and an OT sergeant at full strength. It has a modular hospital tent to provide a 20-bed hospital ward. The MF2K hospitals were designed to be at one location, but may follow the MRI doctrine of splitting off a portion to a remote site, taking with it some of the psychiatry service personnel. One portion would not have the psychiatry ward module, and would need to function similarly to the MRI CSH psychiatry service.

9-12. The MRI CSH has a psychiatrist, a psychiatric nurse, and two MH specialists. The service does not have an organic psychiatric ward where NP patients are admitted. There is normally an intermediate care ward with medical ward staff designated to receive NP patients. Some of these patients seen by the NP clinic will already be stable, and are being evaluated and prepared for RTD or evacuation. Most patients who are admitted for emergency or full stabilization will be on the designated ward, although a few with serious medical complications could require the intensive care ward. If the hospital has a high census of medical and surgical patients, most of the NP patients will need to occupy beds on the intermediate care ward. The MRI hospital may be augmented by staff from a medical company/detachment, CSC to operate a NP ward, as required.

9-13. Full stabilization facilities in theater can be categorized into two types:

- Mobile facilities use general purpose (GP) large or medium tents as used in medical units, CSC and can also be available to a CSH or a tent, expandable, modular, personnel (TEMPER) tent as used in a CSH.
- Fixed facilities use buildings that were previously hospitals or buildings converted to hospitals.

9-14. Facilities used for full stabilization include tents, Deployable Medical Systems (DEPMEDS) TEMPER tents, and fixed facilities. The adaptation of these facilities have both advantages and disadvantages that include—

- The principal advantage of the (hospital) TEMPER tents, as assembled into DEPMEDS hospital, is their climate control capability. This may be a significant safety advantage for treating seriously disturbed patients in restraints with high-dose medication, which can disrupt body's ability to regulate normal body temperature. The TEMPER and standard tents both pose greater problems for security than do fixed facilities. The staff may, therefore, have to rely more than is ideal on physical restraints and medications for sedation of some cases. Blankets or screens can be used to isolate or segregate problem patients from others. Such partitions reduce mental contagion but provide little true protection.

- Standard hospital beds are on high, lightweight metal legs with wheels. For full stabilization purposes, these should be replaced with standard low, stable cots to hold strong, agitated patients in restraints. The cots also make a more "military" setting and can be used as seats for group activities.
- When feasible, it is best to have a separate "closed" (high security) and "open" (moderate/minimal) security area. The latter could be a standard GP large tent (the same as those of the minimal care wards [MCWs] located close to the TEMPER tent of the official NP ward). The specialists (MOS 68WM6 [practical nurse] and 68W [health care specialist]) of the MCW could be given on-the-job training in supervision and military group activities for the moderate/minimum security cases if the NP staff is too small.

Chapter 10

Combat and Operational Stress Control Soldier Restoration

SECTION I — SOLDIER RESTORATION

SOLDIER RESTORATION PROGRAM

10-1. Soldier restoration is normally a 24- to 72-hour (1- to 3-day) program in which Soldiers with COSR receive treatment. Soldier restoration is accomplished using the principles of BICEPS and the 5 R's as discussed in the previous chapter. The 5 R's are tailored to the needs of the Soldier. Soldier restoration is conducted as close to the Soldier's unit as possible. Soldier restoration can be conducted by medical units/elements throughout the theater with the assistance of organic and/or augmented COSC personnel. The medical company/detachment, CSC is staffed and equipped to establish Soldier restoration programs. Soldier restoration may also be accomplished at a Level III MTF (CSH). The tenets remain the same but implementation may differ in the duration of Soldier restoration and the specialized skill and knowledge of available providers. There are three subcategories of Soldier restoration (referred to as lines of Soldier restoration) based on location and available resources that are discussed in Section II of this chapter.

SOLDIER RESTORATION PROCESS

10-2. The process of Soldier restoration involves several steps that include screening, assessment, and interventions, reintegration/coordination, and movement of Soldiers. A discussion of each of the steps in the process is provided below.

SCREENING

10-3. Adequate medical screening and treatment as necessary must be done before the Soldier begins the Soldier restoration program. Minor medical conditions can be treated during routine sick call. Soldiers entering a Soldier restoration program should be only those *hold* cases that require continuous medical and/or BH evaluation and observation for 24 hours or more. The criteria for *hold* cases are discussed under COSC triage in Chapter 8. Tending to and restoring physiological status (such as sleep and hydration) is a priority. Some Soldiers who need Soldier restoration will also have a mental disorder. Treatment for the mental disorder may continue or be initiated during the Soldier restoration process.

ASSESSMENT AND INTERVENTION

10-4. Initial assessment and subsequent COSC interventions depends upon the severity of the COSR and to what extent the reactions interfere with the Soldier's ability to function. More thorough interviews are conducted only after the Soldier's physiological status has been restored.

REINTERGRATION

10-5. Coordination is required to assist Soldiers with reintegration back into their units. Soldiers are reassessed regularly. When a Soldier experiencing COSR begins to improve, the COSC interventions shift toward the reintegration of the Soldier back into his unit. The COSC personnel must work with the COSC consultant of their area or other resources to assist with this reintegration.

MOVEMENT

10-6. Movement of Soldiers from one line of Soldier restoration to another that has greater capabilities or security may be required. When movement of these Soldiers is required, vehicles used to transport them should be nonambulance vehicles, if possible, and accompanied by escorts from their unit. If nonmedical unit members are not available to perform escort duties, medical augmentation may be required.

PRINCIPLES AND PROCEDURES OF SOLDIER RESTORATION

10-7. Initial Soldier restoration begins as close to the Soldier's unit as possible, normally near an MTF where the Soldier can get a reprieve from extreme stress but at the same time can be close to his unit. Normally, Soldier restoration facilities are established in the BSA near or adjacent to the FSMC/BSMC Level II MTF. Soldier restoration is not feasible at locations that are consistently under artillery, air, or direct-fire attack, but ideally are still within the sound of the artillery or other reminders of battle. If there is potential for attack, there must be cover and defensive positions. The location should not be one from which a move is likely within 24 hours. If there is a significant possibility of a move, only those cases that can participate actively in the move with minimal supervision should be managed at this location. The specific site of the Soldier restoration facility should be adjacent to the medical company's area and be out of the immediate (close) sight of the medical triage and treatment areas

REASSURANCE

10-8. Immediate reassurance is given to the Soldier with COSR beginning with the COSC triage as discussed in Chapter 8. Tell the Soldiers that they are temporarily joining the unit, not as patients, but as Soldiers who need a few days to recover from COSR. Emphasize that COSR is a normal response to extremely abnormal conditions. Rapid recovery is also normal and RTD is expected. Reassure the Soldiers about safety and what to do in the event of an attack or march order. Lastly, orient the Soldiers to the Soldier restoration program.

STRUCTURED MILITARY ENVIRONMENT

10-9. An emphasis should be placed on maintaining a highly structured military unit environment and schedule of activities in order to keep the Soldier from adopting a patient role. Assign the Soldier to a squad under supervision of a specific squad leader. The squad leader may be a CSC unit/MH section member or a member of the BSB subordinate unit. The squad leader may also be one of the Soldiers there for Soldier restoration, if their condition allows. The newly arrived Soldier is assigned to a tent that has been designated/erected for the Soldier restoration program. Soldiers with more dramatic COSR or BH symptoms should be temporarily quartered separately from other Soldiers receiving Soldier restoration. In remaining consistent with the principle of treating Soldiers with COSR as Soldiers instead of patients, it is recommended that Soldier restoration facilities do not display the Red Cross as displayed on the MTF. This could affect their protection under the Geneva Conventions. See FM 4-02 for information on the Geneva Conventions.

SUPPORT THE SOLDIER'S MILITARY IDENTITY

10-10. Maintain appropriate rank distinctions, titles, and military courtesies within the confines of the tactical situation. Expect the Soldier to maintain military bearing and personal appearance. They should be in duty uniform. Conduct basic Soldiering skills. Do not take personal possessions away from the Soldier. This includes weapons, unless there is significant concern for the Soldier's safety. If the Soldier arrives with a weapon, SOPs of the AO will guide whether or not the weapon is secure for the Soldier. Also, encourage the unit to maintain contact with its Soldier.

REPLENISHMENT OF PHYSIOLOGICAL STATUS

10-11. Get the Soldier under shelter and cool down if overheated, warm up if cold, and dry off if wet. Providing hot beverages and/or soup will also assist with restoring body temperature. Replenish hydration with palatable beverages and meals. Unless the Soldier is totally exhausted, institute some personal hygiene. If the Soldier needs uniforms or equipment, coordinate with the BSB S4 or supporting logistic element. Restorative sleep should be as normal as possible. Soldiers are typically able to fall asleep when reassured of safety, and do not routinely require medication to induce sleep. They should be informed that they may awaken with vivid and frightening dreams, and be instructed on quick relaxation techniques to go back to sleep. If those measures are insufficient, one-time medication with a sedative/hypnotic may be considered, coordinated, and prescribed by a physician or PA from the supporting medical company. The duration of sleep should be sufficient to make substantial progress in repaying the sleep debt. It should also begin the process of restoring a reasonable sleep/wake cycle that is consistent with the Soldier's duties in his unit.

RESTORE CONFIDENCE

10-12. Restoring a Soldier's confidence may include—

- Providing regularly scheduled formations to keep Soldiers informed of daily activities and the tactical situation, including information about their unit, when available.
- Providing therapeutic occupations that are based on the Soldier's current functional ability. The OT (or COSC team members under the guidance of the OT) selects therapeutic occupations that support the Soldier's military identity and enhance the Soldier's sense of competence. Therapeutic occupations may include activities of daily living (such as physical fitness or uniform maintenance); educational activities (selected common and collective Soldier task training or life skills training); work/productive activities (militarily relevant tasks such as vehicle maintenance or site maintenance); leisure/recreational activities (enjoyable, relaxing activities, games, and sports); social participatory activities (cooperative/competitive sports, games, ceremonies, or celebrations). Therapeutic occupations provide the casualty with a challenge but afford successful performance that shows the Soldier that he is still capable and competent. This realization plays the dominant role in restoring the Soldier's sense of confidence, functional capacity, and ability to RTD.
- Training and teaching Soldier on methods for managing excessive stress.
- Facilitating factional review of precipitating event that has caused the Soldier's COSR through talking (ventilating) and coaching. Combat and operational stress control personnel (or health care providers trained by COSC personnel) facilitate these discussions. They help Soldier restore perspective with questions and coaching. The factual review (debriefing) is often best done individually unless several Soldiers experienced the same event.

SECTION II — LINES OF SOLDIER RESTORATION

FIRST-LINE SOLDIER RESTORATION

10-13. First-line Soldier restoration is usually provided at the BSMC/FSMC in the BSA. It is provided by personnel organic to the BSMC/FSMC usually assisted and supervised by COSC personnel organic to the medical company and/or from CSC unit personnel. This also applies to a cavalry regiment. Because of the high mobility of the BSMC/FSMC, Soldier restoration will often be a 24- to 72-hour process. First-line Soldier restoration also may occur in relative secure locations in division, corps, and theater AOs. First-line Soldier restoration in the division is provided near and supported by a Level II MTF such as a MSMC with its organic MH and/or minimal augmentation from a CSC unit. First-line Soldier restoration in corps and theater is provided by ASMCs with responsibility for their AO. It is provided by COSC personnel organic to the medical company and/or from CSC unit personnel. The MSMC and ASMCs will be relatively unlikely to move on short notice so Soldier restoration up to 72 hours should be feasible. Support provided for Soldier restoration is a responsibility of the MH sections of supporting medical companies. Tasking for support of Soldier restoration programs is accomplished through the command surgeon.

SECOND-LINE SOLDIER RESTORATION (SOLDIER RESTORATION CENTER)

10-14. Each of the first-line Soldier restoration locations listed above should be backed up by a second-line Soldier restoration capability at a location that is less likely to have to move on short notice and has more COSC capabilities. Soldier restoration centers may be located near the MSMC, ASMC, or with a CSC medical company or detachment. They can be located in the BSA near the BSMC/FSMC when it is at a stable base camp. The Soldier restoration center may receive *hold* cases that are transferred from the first-line medical companies, as well as Soldiers from nearby units. Second-line Soldier restoration has more equipment and a greater range of COSC expertise. This permits a 72-hour holding capacity for a stable well-organized Soldier restoration center and may provide full stabilization.

THIRD-LINE SOLDIER RESTORATION

10-15. In some scenarios, units with Soldiers in need of Soldier restoration may be significantly closer to a Level III MTF such as a CSH than to a medical company or CSC unit. On order, a Soldier restoration program may be conducted by the CSH specialty clinic NP staff, which may also be augmented with personnel from a medical company/medical detachment, CSC. In such cases, a Soldier restoration program may be conducted by the CSH specialty clinic NP staff. It may be augmented with personnel from the medical company or detachment, CSC. Soldier restoration near a CSH must be kept clearly separate from the patient ward. It may be done at a MCW or in separate tents.

RETURN TO DUTY OF RECOVERED COMBAT AND OPERATIONAL STRESS REACTION SOLDIERS

10-16. Most COSR symptoms do not necessarily improve completely while the prospect of combat continues. The Soldiers should be given the positive expectation that they will RTD. Every possible effort should be made to return Soldiers to their original unit. The RTD of Soldiers following Soldier restoration depends on how near the Soldier's unit is, the availably of a means of transportation, and the tactical situation. Ideally, their units are contacted to send someone to get the Soldier or he may be returned to his unit by way of the personnel replacement company. Mental health personnel coordinate directly with the unit to which the Soldier is returning and/or with COSC, other medical personnel, or UMT supporting the unit's AO. These contacts can consult with the leaders of the Soldier's unit and facilitate the Soldier's acceptance back into his unit.

DOCUMENTATION

10-17. Soldiers receiving Soldier restoration interventions must be tracked from the initial contact until they are returned to their unit. A record must be maintained of interventions and activities provided during Soldier restoration, as well as the Soldier's response. A notation of the dates, any pertinent medical data, and providing unit is entered in the Soldier's medical record. An administrative summary of Soldier restoration services may be developed, stored, and disposed of as directed by the Soldier restoration center SOP. If the Soldier received BH treatment, documentation should be maintained as delineated in AR 40-66. A statistical record is maintained and sent to higher medical headquarters according to SOP.

Chapter 11

Behavioral Health Treatment

SECTION I — BEHAVIORAL HEALTH CARE

TREATMENT FOR BEHAVIORAL DISORDERS

11-1. Behavioral health treatment exists when there is an explicit therapist-patient or therapist-client relationship. Behavioral health treatment is provided for Soldiers with behavioral disorders to sustain them on duty or to stabilize them for referral/transfer. This is usually brief, time-limited treatment as dictated by the operational situation. Behavioral health treatment includes counseling, psychotherapy, behavior therapy, occupational therapy, and medication therapy. Treatment assumes an ongoing process of evaluation, and may include assessment modalities such as psychometric testing, neuropsychological testing, laboratory and radiological examination, and COSC providers' discipline-specific evaluations.

COMBAT AND OPERATIONAL STRESS CONTROL BEHAVIORAL HEALTH PROVIDER

11-2. There are five professional disciplines and two enlisted specialties that serve as BH/COSC providers, see Chapter 3. Army BH/COSC providers are officers who hold active individual clinical privileges granted by an MTF according to AR 40-68. Providers also include enlisted medical specialist who do not hold individual clinical privileges, but who work under direct supervision of a professional COSC provider. Behavioral health treatment is provided within the scope of the providers' clinical privilege, training, and experience.

SECTION II — BEHAVIORAL HEALTH TREATMENT PROTOCOLS AND MEDICATIONS

BEHAVIORAL HEALTH TREATMENT FOR SOLDIERS

11-3. Behavioral health treatment is provided to Soldiers with diagnosed behavioral disorders (see Chapter 8), and who require more intentions for their diagnoses. It is both inappropriate and detrimental to treat Soldiers with COSR as if they are a BDP. A therapeutic relationship may promote dependency and foster the "patient" role. Likewise, medication therapy and the highly structured treatment modalities imply the "patient" role. Medication for transient symptom relief (insomnia or extreme anxiety) may not be detrimental if there is no expectation that medication will continue to be prescribed.

BEHAVIORAL HEALTH PROTOCOLS AND PROCEDURES FOR TREATMENT

11-4. Behavioral health treatment is provided within a context of preventing symptoms of behavioral disorder from progressing in severity and improving the level of functioning. The principles of BICEPS apply. Treatment should not impede activities for other prevention levels (for example, universal, selective, and indicated). Treatment should not interfere with the Soldier's duties or the unit's mission. The COSC provider should emphasize Soldier's identity as a Soldier despite having a diagnosis. Transportation of the Soldier to the provider (or vice versa) and ability to assure reliable meeting times can be limiting factors, depending on the operational environment. Medication refills must be coordinated for availability.

Standards of Treatment

11-5. Treatment standards are the same in the deployed environment as in garrison. When operational requirements dictate that clinical standards of treatment/care are waived or relaxed, it must be approved by the AO COSC consultant. (AR 40-68 provides additional information.) Treatment should be tailored to the anticipated availability of the Soldier and COSC provider. Short-term interventions are more practical than long-term commitments. If longer-term treatment is necessary, design the intervention in time-limited modules. Under no circumstances should treatment diminish the Soldier's ability to provide self-care and to defend himself. Exceptions include emergency stabilization and preparation for evacuation. In addition, the VHA/DOD Clinical Practice Guidelines website (http://www.oqp.med.va.gov/cpg/cpg.htm) offer clinicians evidence-based assessment and treatment algorithms for acute stress disorder, PTSD, and many other behavioral disorders.

Army Regulations Governing Evaluations

11-6. Fitness for duty evaluations are conducted as necessary within the priorities of the supported commanders according to AR 40-501; psychiatrists should not initiate a medical evaluation board without first ensuring the Soldier has received adequate treatment. This treatment may not be available in theater. Command-directed evaluations are conducted as necessary within the priorities of the supported commanders according to DODD 6490.1 and MEDCOM Regulation 40-38. Clinical documentation should be safeguarded according to AR 40-66 and local command policy. Treatment should be conducted in a location that is as private as possible. Information can be released to a third party if the Soldier consents. Combat and operational stress control providers need to notify command when the Soldier's safety is in question (suicidal, homicidal) or if the Soldier is removed from his unit for medical observation. Additional release of information to command is on a need-to-know basis. For any questions on release of information on COSR and NP patients, consultation with supporting JAG office is advised. Also, maintaining the Soldiers health record with clinical data is required and accomplished according to AR 40-66 and supplemental theater policy as appropriate. Treatment should occur throughout the evacuation process and follow-up is expected at home station.

Chapter 12
Reconditioning

SECTION I — RECONDITIONING PROGRAM

LOCATION

12-1. Reconditioning programs are intensive efforts to restore those Soldiers triaged as a *refer* case, but who still have good potential for RTD. Referral to reconditioning can be from Level II MTF, COSC Soldier restoration program, or from Level III MTF. Reconditioning includes the rehabilitation of Soldiers with mental disorders, such as substance abuse/dependency. Reconditioning programs are conducted by COSC personnel and Soldiers usually participate for up to 7 days. Soldier participation may be extended by a case-by-case exception to theater evacuation policy. Reconditioning is conducted only in a theater where there are adequate COSC elements and supplies. When COSC resources are needed for other functional areas, any ongoing reconditioning program reduces its scope of services or closes. Reconditioning is provided at first-, second-, or third-line centers determined by location and available resources.

PROVIDERS

12-2. Conducting reconditioning programs is a mission of the medical company, CSC and the medical detachment, CSC (MRI). Reconditioning can also be done on a small-scale by a CSC medical detachment (MF2K) under some circumstances.

SECTION II — TENETS AND PROCEDURES OF RECONDITIONING

RECONDITIONING

12-3. Reconditioning may be considered an extension of Soldier restoration. Reconditioning is similar to Soldier restoration, but with potentially longer stay, treatment strategies focus on preventing atrophy of skills and assisting Soldiers in regaining skills needed for duty. Reconditioning makes more use of BH treatment modalities.

STRUCTURED MILITARY ENVIRONMENT TO SUSTAIN SOLDIER'S IDENTITY AS A SOLDIER

12-4. Like Soldier restoration, reconditioning emphasizes a highly structured military unit environment and schedule of activities in order to keep the Soldier from adopting a "patient" role. Maintaining a military environment is even more critical when colocated with hospitals or other service support elements.

REPLENISHMENT OF PHYSIOLOGIC STATUS AND CONFIDENCE

12-5. Reconditioning initially emphasizes physical replenishment and hygiene, but later shifts the emphasis to more closely match the conditions that the Soldier should expect when RTD. In order to restore confidence in the Soldier, unit formations are held on a regularly scheduled basis, and include the Soldiers receiving reconditioning.

OCCUPATIONAL THERAPY

12-6. Therapeutic occupations may include—

- Activities of daily living (for example, hygiene, physical fitness, and uniform maintenance).

- Educational activities (for example, common and collective Soldier task training, and life skills training).
- Militarily relevant work details according to the current functional level and MOS/duties of the Soldier.
- Leisure/recreational activities (for example, enjoyable, relaxing activities, including cooperative physical and mental activities, and basic relaxation techniques).
- Social participatory activities (for example, games and ceremonies).
- Group training in relaxation techniques.
- Ventilation and coaching where COSC (or medics trained by COSC personnel) encourage discussion about stressors and their impact on mental well-being. This technique helps the Soldier to restore personal perspective with questions and coaching. Individual counseling and therapy may improve the Soldier's functions as well.

12-7. When reclassification recommendations are considered, an adjacent CS or CSS unit may be able to provide a job that will match the Soldier's abilities. This provides an opportunity to demonstrate the Soldier's abilities and build confidence. The reconditioning personnel may recommend to S1/G1 that the Soldier be reclassified to another MOS.

FIRST-LINE RECONDITIONING PROGRAM

12-8. First-line reconditioning programs in the corps are staffed by task-organized CSC elements from the CSC company or by an MRI CSC detachment. Reconditioning can be conducted by an MF2K CSC detachment, but would preempt its Soldier restoration capability. If the inpatient NP workload is light, a small reconditioning center may be staffed by personnel from the NP ward and consultation service of an MF2K CSH. The first-line reconditioning center usually collocates with a CSH but must maintain its separate, nonhospital identity. It should not be situated among the hospital wards, or near the morgue, triage area, or helicopter pad. Within a theater/AO, the preferred option is to have one or two reconditioning centers placed such as to allow easy access from Soldier restoration centers. Under major, prolonged combat conditions, it may be more appropriate to have one reconditioning center behind each division. The reconditioning center is dependent on the hospital for support. The reconditioning center works with the supporting hospital by sending work parties of Soldiers in the program to assist in food preparation and delivery and cleanup chores. These Soldiers may also be used for assisting with work details throughout the hospital, but must be under direct supervision of either hospital or BH personnel. The reconditioning center uses the supporting hospital's medical records section to maintain the permanent case records. Cases in the reconditioning center are counted as patients in the reconditioning center on the daily hospital census. The cases are not counted as occupied beds when reporting the hospital bed occupancy. Upon disposition from the reconditioning center, whether for RTD, retraining for other duty, or evacuation, the reconditioning center prepares the chart for further evacuation or writes the discharge summary and closes the hospital's chart.

12-9. Some reconditioning cases will be able to return to far forward CS or CSS duty. However, many of the Soldiers who need reconditioning will be unable to return to their original unit, due to combat operations or the nature of their symptoms. Soldiers who undergo a 4- to 7-day Soldier restoration program in the corps should not be crossed off the division's personnel rolls (as specified in AR 40-216). Continuing contact with the unit increases RTD rates. For Soldiers who recover but are no longer on the division's rolls, every feasible effort should be made to return them to their original units. The chief, division MH and the CSC unit teams that are attached to reinforce the division should maintain frequent contact with the reconditioning centers that support the division. Coordination efforts, through the division support command (DISCOM) sustainment brigade and the personnel replacement system by division BH personnel, should attempt to facilitate return of recovered Soldiers to their original units. If recovered Soldiers cannot be returned to their previous small unit, consider forming them into cohesive pairs or small groups which can be reassigned to a new unit together. The standard corps evacuation policy is seven days, but some Soldiers with good potential for RTD may need a few more days at the reconditioning center. The corps evacuation policy is at the discretion of the theater commander. The principle purpose of a short evacuation policy is to keep hospital beds available for mass casualties and to minimize the expense and labor required to treat serious surgical and medical cases.

SECTION III — EVACUATION POLICY

EXTENDING THE THEATER EVACUATION POLICY

12-10. If need assessment indicates that a longer period of time (more than seven days) will achieve a Soldiers' RTD, the theater COSC consultant can request permission to extend the theater evacuation policy (up to 14 days). He submits the request for authority via the commander, MEDBDE, through the corps surgeon to the corps commander. The following facts apply: The reconditioning program, because of its austerity, is not a significant logistical burden to the corps. Soldiers in reconditioning perform useful work details and perimeter defense. Lastly, the increase in RTD is important in operations such as stability and reconstruction operations, where reducing personnel attrition becomes an important factor.

EVACUATION

12-11. Reconditioning cases that do not recover sufficiently to return to some duty within the designated evacuation period are evacuated from corps to EAC. They are best transported in cargo trucks and buses, bus ambulances, or an ambulance train. In the latter two cases, they should be assigned helper tasks. Use air evacuation only if there is no other alternative.

SECOND-LINE RECONDITIONING

12-12. Second-line reconditioning is conducted in the intermediate staging base outside the CZ in the theater. This center could be at a CSH, a fixed MTF, or a CSC unit element. The second line reconditioning center continues to emphasize physical fitness, Soldier skills, work details, and individual/group counseling/psychotherapy. Cases will be retrained for CSS duties at corps and theater levels. As soon as the recovering Soldiers are ready, the retraining site can shift to on-the-job training at a nearby CSS unit.

THIRD-LINE RECONDITIONING

12-13. Third-line reconditioning is for Soldiers with COSR and/or mental disorders who did not improve sufficiently at lower lines of reconditioning but still have RTD potential. These Soldiers are transferred to a reconditioning program at a regional MEDCEN or home station MEDDAC. These reconditioning centers, like the others, must maintain a military atmosphere and provide opportunities to engage in occupational therapy. Some Soldiers may also require retraining for other duty/MOS reclassification.

Appendix A

Combat and Operational Stress Control and Religious Support

ROLE OF UNIT MINISTRY TEAM

A-1. This appendix addresses the general role of the UMT in the commander's program for COSC. The UMT, imbedded within units down to battalion level, provide immediate support to leaders in fulfilling their COSC responsibilities. The UMTs also assist in training leaders to recognize combat and operational stress identification and intervention responsibilities. In cooperation with unit medical personnel, UMTs serve as a primary referral agency to BH resources.

RELIGIOUS SUPPORT FOR COMBAT AND OPERATIONAL STRESS CONTROL

A-2. Soldiers' inner resources are generally rooted in their religious and spiritual values. In combat, Soldiers often show more interest in their religious beliefs. When religious and spiritual values are challenged by the chaos of combat, Soldiers may lose connection with the inner resources that have sustained them. The UMT is the primary resource available to Soldiers experiencing such dilemmas and is a valuable resource in assisting them as they seek to refocus their spiritual values.

Unit Ministry Team Support for Combat and Operational Stress Control

A-3. The UMT provides preventive, immediate, and restorative spiritual, emotional support and care to Soldiers experiencing COSR.

Preventive Religious Support for Combat and Operational Stress Reaction

A-4. The UMT assists in preventing COSR and misconduct stress disorders through spiritual fitness training. Ministry of presence with Soldiers, assigned Department of Army civilians, and contractors is critical. The UMT provides a stabilizing influence on personnel and assists them in strengthening and regaining personally held spiritual values. Preventative activities include—

- Worship opportunities.
- Private and group prayer opportunities.
- Religious literature and materials.
- Scripture readings.
- Sacraments and ordinances.
- Assistance to personnel and families prior to deployment, emphasizing family strengths.

Immediate Religious Support for Combat and Operational Stress Reaction

A-5. The UMT assists commanders in the identification of personnel experiencing negative reactions to combat and operation stress, COSR, and misconduct stress behaviors. The UMT works closely with the unit's leaders and medical personnel to care for COSR cases through religious support and comfort. Immediate religious support activities may include—

- Conversation focused upon fears, hopes, and other feelings.
- Prayer for fallen comrades and memorial ceremonies and services.
- Rites, sacraments, and ordinances, as appropriate.
- Sacred scripture.

Restorative Religious Support for Combat and Operational Stress Control

A-6. Following an operation, a unit may require reconstitution. Surviving Soldiers may need to rebuild emotional, psychological, and spiritual strength. Depending upon the spiritual, emotional, and physical condition of the unit's Soldiers, the organic UMT may need augmentation from higher echelons or other units. Restorative religious support activities may include—

- Worship, sacraments, rites, and ordinances.
- Memorial ceremonies and services.
- Religious literature and materials.
- Grief facilitation and counseling.
- Reinforcement of the Soldiers' faith and hope.
- Opportunities for Soldiers to talk about combat experiences and to integrate those experiences into their lives.

Appendix B

Medical Detachment, Combat and Operational Stress Control

CONCEPT OVERVIEW

BACKGROUND

B-1. The current medical units, CSC in the Army inventory were initially developed over ten years ago. The MF2K redesign resulted in the development of the MRI. Recent events in the Global War on Terrorism in both Afghanistan and Iraq have indicated a need to update these units from MRI medical detachment, CSC to a multifunctional COSC detachment. The 2003 US Army Surgeon General's Mental Health Advisory Team (MHAT) report and observations indicate a need to update the unit in terms of modularity, multifunctionality, and its working relationship with support units and command surgeons from brigade to theater levels.

PROPOSED MEDICAL DETACHMENT, COMBAT AND OPERATIONAL STRESS CONTROL

B-2. The proposed medical detachment, COSC (TOE 08660G000) consists of a detachment headquarters, a main support section, and a forward support section. The main support section consists of its headquarters and an 18-Soldier BH team made up of social workers, clinical psychologists, psychiatrists, occupational therapists, psychiatric nurses, MH specialists, and OT specialists. The forward support section consists of an 18-Soldier BH team only. Each BH team is capable of breaking into six 3-person subteams, for battalion/company prevention and fitness support activities. This provides for a total of 12 subteams for each detachment, giving supported commanders more teams and more flexibility in the utilization of those teams, yet maintaining all of the capabilities of COSC.

Operational Environment

B-3. Soldiers will continue to experience the physical and mental impacts of high stress in both combat and stability and reconstruction operational environments. Advances in technology which will increase effectiveness, and the impact of the individual Soldier, will also increase dispersion and add to isolation and stress levels of Soldiers. The enemy of the future will look for methods that will have the greatest psychological impact on our future Soldiers; future enemies will not be our technological equals and will increasingly rely on terror and nontraditional methods to unnerve, injure, and demoralize both Soldiers and civilians. There is an identified need to update COSC units in terms of flexibility, modularity, multifunctionality, and its relationship to supported units and the surgeons responsible to commanders for the BH mission. The current configuration of the COSC units resulted in an uneven use of available COSC resources and that there was a greater need for preventive and limited fitness services forward. Commanders have communicated there is a greater need for active outreach and consultation.

Assumptions

B-4. Power projection will likely remain the fundamental strategic and operational imperative of our forces for the foreseeable future. The new medical detachment, COSC will be well suited to the future force, designed to move with and support the division and corps of the future force, and division and corps equivalents. The medical detachment, COSC will be assigned to a MEDBDE, MMB, or MEDCOM and will be operationally attached to supported units.

B-5. Each BCT has a MH section with one BH officer and one MH specialist and that all division and brigade MH sections are filled. The medical detachments, COSC will be able to augment BCTs MH sections. The medical detachment, COSC will require additional logistical, finance, maintenance, personnel, legal, FHP, and administrative services support.

Implication for Combat and Operational Stress Control

B-6. Future forces must be emotionally and mentally fit before deploying and resilient in battle in order to endure the fast pace and intense lethality of the battlefield. Commanders will need to be able to apply COSC capabilities at every level quickly and effectively. Reorganization of the medical detachment, CSC to a combined, multifunctional medical detachment, COSC will provide greater flexibility and greatly increase outreach and preventive services while maintaining capability for the entire spectrum of COSC.

Operational Concept, Mission

B-7. The medical detachment, COSC may be attached to a theater or division medical C2 headquarters in order to provide COSC casualty prevention, treatment, and management on an area basis.

B-8. The AUTL (see FM 7-15) linkage includes—

- ART 6.5. Provide Force Health Protection in a Global Environment.
- ART 6.5.1. Provide Combat Casualty Care.
- ART 6.5.1.5. Provide Mental Health/Neuropsychiatric treatment.
- ART 6.5.4. Provide Casualty Prevention.
- ART 6.5.4.5. Provide Combat Operational Stress Control Prevention.

Method of Employment

B-9. The medical detachment, COSC deploys with a division-sized force into a theater or with a task force of up to 25,000 Soldiers in stability and reconstruction operations. The medical detachment, COSC is attached for OPCON to the division/corps/theater medical C2 headquarters. The medical detachment, COSC is assigned to the MMB of units that have reorganized under the future force design. It is normally tasked by the command surgeon to support units within the command. The detachment commander as authorized by his medical C2 headquarters will coordinate operations with the command surgeon or the division/corps/theater psychiatrist. In the absence of a psychiatrist, the detachment commander will advise and consult with the command surgeon and the commander as required. Subteams may be task-organized and further attached for operational control to maneuver BCTs, and will operate in and forward from BSA, working closely with medical assets and combat religious support team (CRST). The detachment commander will coordinate efforts with the division and the CRST to provide in-theater prevention programs as required (suicide prevention, reunion, COSC training, TEM, and PTE stress management).

Detachment Headquarters

B-10. Plans, supervises and monitors the activities of the medical detachment, COSC in support of the COSC mission. See Table B-1 for a listing of personnel assigned to the detachment headquarters. The commander will establish a command post (CP) collocated with the supported unit surgeon (either division or sustainment brigade) and coordinate all activities of the detachment through the surgeon section. The detachment commander provides C2 and works closely with supported unit surgeon to provide logistics and other required services for the detachment. The field medical assistant and/or the detachment sergeant also establishes and coordinates logistics and required services support through the supported command staff for all detachment personnel and operations.

Table B-1. Detachment headquarters personnel

Detachment Commander (LTC/05, AOC 05A00)
Medical Operations Officer (CPT/03, AOC 70B67, MS)
Chaplain (CPT/03, AOC 56A00, CH)
Detachment Noncommissioned Officer in Charge (NCOIC) (SFC/E7, MOS 68X40)
Supply Sergeant (SGT/E5, MOS 92Y20)
Human Resources Specialist (SPC/E4, MOS 42A10)
Wheeled Vehicle Mechanic (SPC/E4, MOS 63B10)
Patient Administration Specialist (SPC/E4, MOS 68G10)
Religious Support Specialist (SPC/E4, MOS 56M10)
Cook (PFC/E3, MOS 92G10)
Note. The position of detachment commander may be filled by a (LTC, AOCs 73A00, 73B00, 60W00, 65A00 or 66C00). For description of duties, see Section I of the TOE and Department of Army (DA) Pamphlet 611-21.

Main Support Section

B-11. Included in the main support section is the detachment headquarters personnel identified in Table B-1 above and an 18-personnel BH team. See Table B-2 for list of personnel assigned to the BH team. This BH team can break into 6 three-person subteams, each with one officer, one NCO, and one specialist to conduct prevention and limited fitness operations in the battalion/company level in a maneuver BSA. Teams are collocated with a BCT medical unit. Modularized subteams may be combined and task-organized to provide or support any other COSC function, which includes larger scale Soldier restoration reconstitution, or any other COSC or BH mission as determined by its headquarters. Tasks for the main support section BH team includes:

- Consultation and coordination with commanders at team level (battalion and company).
- Coordination with CRST.
- Prevention and fitness support activities.
- Reconstitute any other COSC function as required.

B-12. When attached for support to any medical company Level II MTF, the grouped subteams are dependent on the supporting medical company for logistical support and quarters. The subteam can provide 3-day restoration for up to 50 Soldiers. With additional logistical support and quarters from supported command, the team can accommodate additional Soldiers under surge conditions.

Table B-2. Main section behavioral health team

Team Leader/Psychiatrist (MAJ/04, AOC 60W00)
Occupational Therapist (MAJ/04, AOC 65A00)
Social Work Officer (MAJ/04, AOC 73A67)
Behavioral Science Officer (CPT/03, AOC 67D00)
Clinical Psychologist (CPT/03, 73B67)
Psychiatric Nurse (CPT/03, 66C7T)
Occupational Therapy NCO (Four)—(Two SSGs/E6 and Two SGTs/E5, MOS 68W30/68W20)
Mental Health NCO (Four)—(One SSG/E6 and Three SGTs/E5, MOS 68X30/68X20)
Mental Health Specialist (Six)—(SPC/E4, MOS 68X10)
Note. For description of duties, see Section I of the TOE and DA Pamphlet 611-21.

Forward Support Section

B-13. This section consists of an 18-Soldier BH team. It is comprised of a section leader and five other BH officers, six MH/OT NCOs, and six MH specialists. This BH team can break into six three-person subteams, each with one officer, one NCO, and one specialist to conduct prevention and limited fitness operations in the BCT medical company in the BSA. Modularized subteams may be combined and task-organized to provide or support any other COSC function, which includes larger scale Soldier reconstitution, or any other COSC or BH mission as determined by its headquarters. The forward support section is dependent on the supported unit for C2 and logistical support. See Table B-3 for list of personnel assigned to the forward support section BH team.

Table B-3. Forward support section behavioral health team

Team Leader/Psychiatrist (MAJ/04, AOC 60W00)
Psychiatric Nurse (MAJ/04, 66C7T)
Clinical Psychologist (MAJ/04, 73B67)
Occupational Therapist (CPT/03, AOC 65A00)
Social Work Officer (CPT/03, AOC 73A67)
Behavioral Science Officer (CPT/03, AOC 67D00)
Occupational Therapy NCO (Two)—(One SSG/E6 and One SGT/E5, MOS 68W30/68W20)
Mental Health NCO (Four)—(One SSG/E6 and Three SGTs/E5S, MOS 68X30/68X20)
Mental Health Specialist (Six)—(SPC/E4, MOS 68X10)
Note. For description of duties, see Section I of the TOE and DA Pamphlet 611-21.

Organizational Concept

B-14. Required capabilities require the unit to—

- Deploy to power projection platform or point of debarkation and deploy with supported unit or deploy to theater and join supported unit.
- Provide all functions of COSC from company to theater level in close coordination with and under the direction of the supported unit surgeon.
- Provide prevention, assessment, treatment, and referral BH activities.
- Provide consultation with commanders from company to theater level.
- Conduct and coordinate administrative and logistical support to sustain operations.

Basis of Allocation Total Army Analysis

B-15. The basis of allocation is 0.333 per BCT; 1 per division; 2 per theater. In support of a theater, a COSC detachment provides support on an area basis and provides additional support to the division/corps on order. See Figure B-1 for breakdown of separate standard requirement codes (SRCs) of the medical detachment, COSC sections.

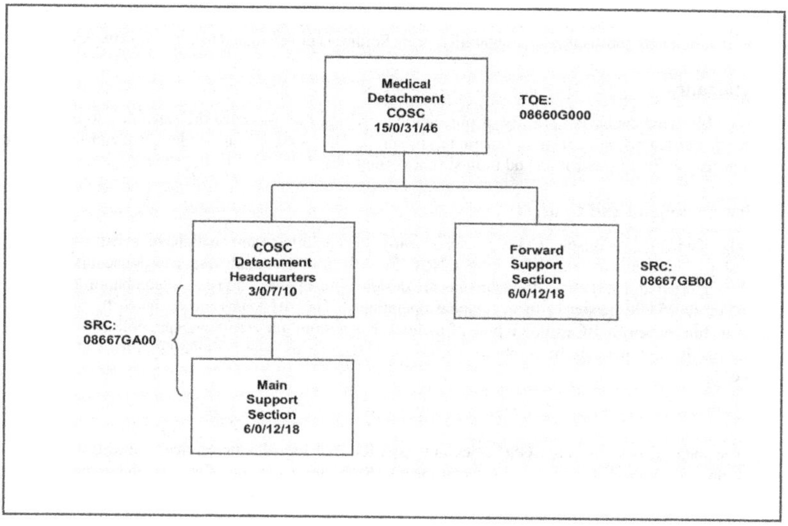

Figure B-1. Medical detachment, combat and operational stress control

Concept of Transition

B-16. Transition begins with the four active component detachments, one at a time. Personnel, transportation, and equipment changes will be minimal; transition should take a short amount of time and money to accomplish. Changes in the detachments are congruent with Force Modernization designs and personnel proponency issues. The proposed COSC detachment is designed with the Future Force in mind.

Patient Medical Records

B-17. The Surgeon General's MHAT report notes that there is a need to standardize BH reporting and documentation. The BH charts were inconsistently maintained and documentation did not always accompany patients through the evacuation chain. An additional finding was that there was no standardized method of collecting BH workload or clinical data and that no database that tracked evacuees provided reliable clinical information. The patient administration specialist (MOS 68G10) will supervise or perform functions as outlined in DA Pamphlet 611-21 for the Soldier caseload; maintain Soldier/patient accountability and records; interface with patient administration division of supported/collocated medical units.

Role of Religious Support Specialist

B-18. Findings of the MHAT include the need for chaplains to be aware of their role in COSC and that an aggressive chaplain outreach program should be executed. This requires that supervisory chaplains be involved in CRST integration with BH and primary care providers, and educate primary players in role of CRST in COSC. These activities require that the COSC chaplain be mobile and have an active presence in the maneuver brigades. This position is required to fulfill the full-time force protection requirements for the chaplain while in theater. The religious support specialist assists the COSC chaplain by screening individuals seeking counseling, coordinates for convoy security when the CRST goes out on site visits; assists chaplain with fund request; develops and coordinates needed religious support projects within the unit; coordinates with detachment for all vehicle services, communication system repairs, detachment training requirements and section accountability. This Soldier is cross-trained to develop skills for COSC.

Multifunctionality

B-19. The MH sections have a priority of prevention activities, but also have the capability to conduct or support any combat operational stress function as required. The COSC teams will be outfitted to a standard to gain the capability of prevention and limited fitness activities.

Modularity, Command and Control

B-20. The medical detachment, COSC attached early to supported units will have a better-integrated relationship. Detachment commanders and NCOICs will collocate with and coordinate all activities through supported unit surgeons. Detachments are designed to support a division in combat or a joint task force of up to 25,000 personnel in noncombat operations. The BH teams are designed to support and operate within maneuver BCTs in support of maneuver battalions for area coverage, collocating in BSA with the BSMC. Teams may be combined for task-organization to meet the entire spectrum of COSC as required.

Limitations

B-21. The medical detachment, COSC does not have its own logistics or services capability and must depend on the supported unit for logistical, legal, FHP, food service, finance, and personnel and administrative services. The COSC teams are small (one or two vehicles) and are dependent upon supported units for convoy and security operations. The logistical support necessary (space, tents, cots, and so forth) to provide Soldier restoration/reconstitution functions must be furnished by the supported element (division, corps, or theater).

Assumptions

B-22. Wartime activities will continue to cause physical and mental impacts of extremely high stress in both direct combat and the support operational environment. Advances in technology will impact the individual Soldier by increasing dispersion, and adding to their isolation and stress levels. The enemy of the future will look for methods that will have the greatest psychological impact on our future Soldier; future enemies will not be our technological equals and will increasingly rely on terror and nontraditional methods to unnerve, injure, and demoralize both Soldiers and civilians. Commanders will need to be able to quickly and effectively apply COSC capabilities at every level of the unit.

Appendix C

Medical Company and Medical Detachment, Combat Stress Control (Medical Force 2000)

SECTION I — MEDICAL COMPANY, COMBAT STRESS CONTROL (TOE 08467L000) (MEDICAL FORCE 2000)

MISSION

C-1. The mission of the AOE medical company, CSC is to provide comprehensive COSC support through directed interventions activities and COSC training for supported corps units. It provides DS to maneuver brigades lacking organic BH officers; augments units with BH assets; and provides area support. The medical company, CSC reconstitutes other COSC assets. The medical company, CSC provides COSC interventions and activities to indigenous populations as directed in stability and reconstruction operations, to include domestic support operations, humanitarian assistance, disaster relief, peace support operations, and detention facility operations. The medical company, CSC provides COSC interventions and activities to units in support of their readiness preparation and throughout their deployment cycle.

ASSIGNMENT

C-2. The AOE medical company, CSC (TOE 08467L000) is assigned to a corps MEDBDE (TOE 08422A100) or corps MEDCOM (TOE 08411A000).

EMPLOYMENT

C-3. Company headquarters usually locates with the corps medical headquarters. The employment of its teams includes their dispersal throughout the corps AO. Teams may be attached to an ASMB; a division or brigade medical company; a CSH; or other corps medical unit headquarters.

CAPABILITIES

C-4. At TOE Level 1, the medical company, CSC includes—

- Dividing the preventive section into six mobile COSC preventive teams.
- Dividing the restoration section into four mobile restoration teams with each team being equipped to hold 40 Soldiers at the same time. With additional logistical support, each team can accommodate additional Soldiers under surge conditions.
- Assisting higher headquarters BH staff with planning and coordination of COSC support, identifying the stress threat and mental and physical stressors, and the implementation of COSC functional areas.

DEPENDENCY

C-5. This unit is dependent on—

- Appropriate elements of the corps for legal; finance; field feeding; personnel and administrative services support; laundry and clothing exchange; mortuary affairs support; and security of EPW, detainee, and US prisoner patients.
- The medical headquarters to which it is assigned/attached for FHP; medical administration; logistics (including MEDLOG); medical regulating of patients; evacuation; coordination for RTD; and unit-level equipment and CE maintenance.

MOBILITY AND SECURITY

C-6. This unit is 100-percent mobile. It requires 100 percent of its organic personnel and equipment be transported in a single lift, using its organic vehicles. Upon relocation, Soldiers being held will require additional transportation. This unit is responsible for perimeter defense of its immediate operational area. However, it is dependent on appropriate elements of the corps for additional security, to include security of convoy operations. Personnel of the company (except the chaplain) are provided weapons for their personal defense and for the defense of their patients and/or held Soldiers.

ORGANIZATION

C-7. The AOE medical company, CSC (Figure C-1) is organized into a headquarters section, a preventive section with six (modular) CSCP, and a restoration section with four (modular) combat stress control restoration (CSCR) teams. The medical company, CSC is only present in the MF2K and is an RC organization that is programmed to be in the Army inventory until the year 2010. Under the MRI a new medical detachment, CSC, TOE 08463A000, will replace this unit.

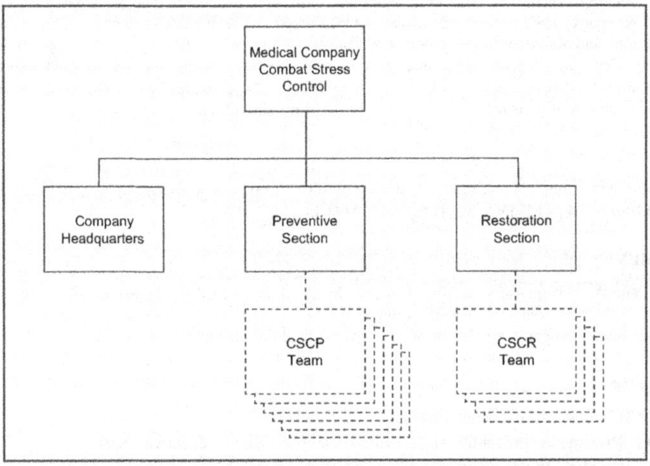

Figure C-1. Medical company, combat stress control

Company Headquarters Section

C-8. The company headquarters section provides C2 and unit-level administrative and maintenance support to its subordinate sections when they are collocated with the company. Personnel from the headquarters section are deployed with teams or task-organized COSC elements, as required. Personnel assigned to this section are identified in Table C-1.

Table C-1. Company headquarters section personnel 3/0/14

Company Commander (LTC/05, AOC 60W00, MC)
Chaplain (CPT/03, AOC 56A00, CH)
Field Medical Assistant (CPT/03, AOC 70B67, MS)
First Sergeant (1SG/E8, MOS 68W5M)
Mental Health NCO (SFC/E7, MOS 68X40)
Motor Sergeant (SSG/E6, MOS 63B30)
Supply Sergeant (SGT/E5, MOS 92Y20)
Equipment Record/Parts Sergeant (SGT/E5, MOS 92A20)
Wheeled Vehicle Mechanic (SGT/E5, MOS 63B20)
Human Resource Specialist (SPC/E4, MOS 42A10)
Nuclear, Biological, and Chemical Specialist (SPC/E4, MOS 74D10)
Wheeled Vehicle Mechanic (SPC/E4, MOS 63B10)
Unit Supply Specialist/Armorer (SPC/E4, MOS 92Y10)
Power-Generation Equipment Repairer (SPC/E4, MOS 52D10)
Cook (Three)— (Two SPC/E4 and One PFC/E3, MOS 92G10)
Note. The cook is deployed to assist the unit providing food service support to the company. He may participate in other CSC unit missions including outreach surveillance and stress control training.

Preventive Section

C-9. The 24-person preventive section staff is identified in Table C-2. The preventive section is task-organized to conduct COSC interventions and activities. Preventive section personnel may be task-organized with personnel of the COSC restoration section into teams for specific missions. The preventive section can divide into six CSCP teams. The section (and team) leader position may be held by any of the officers assigned to the section. The preventive section can augment or reconstitute the medical detachment, CSC teams.

Table C-2. Preventive section personnel 12/0/12

Psychiatrist (Three)—(MAJ/04, AOC 60W00)
Social Worker (Two)—(MAJ/04, AOC 73A67)
Psychiatrist (Three)—(CPT/03, AOC 60W00)
Social Worker (Four)—(CPT/03, AOC 73A67)
Team Chief (Six)—(SGT/E5, MOS 68X20)
Mental Health Specialist (Six)—(SPC/E4, MOS 68X10)

Restoration Section

C-10. The 44-person restoration section staff is identified in Table C-3. The restoration section may be task-organized to perform its COSC mission. Combat stress control restoration section personnel may also be task-organized with personnel of the CSCP section into teams for specific missions. This section can be divided into four CSCR teams. Each CSCR team can deploy a four-person mobile team using their HMMWV.

Table C-3. Restoration section personnel 12/0/32

Occupational Therapist (MAJ/04, AOC 65A00)
Psychiatric/Mental Health Nurse (Two)—(MAJ/04, AOC 66C00/7T)
Clinical Psychologist (MAJ/04, AOC 73B67)
Occupational Therapist (Three)—(CPT/03, AOC 65A00)
Psychiatric/Mental Health Nurse (Two)—(CPT/03, AOC 66C00/7T)
Clinical Psychologist (Three)—(CPT/03, AOC 73B67)
Mental Health NCO (Two)—(SFC/E7, MOS 68X40)
Occupational Therapy NCO (SSG/E6, MOS 68W30/N3)
Team Chief (Four)—(SSG/E6, MOS 68X30)
Patient Administration NCO (SGT/E5, MOS 68G20)
Occupational Therapy Sergeant (Two)—(SGT/E5, MOS 68W20/N3)
Mental Health NCO (Four)—(SGT/E5, MOS 68X20)
Patient Administration Specialist (Three)—(Two SPC/E4 and One PFC/E3, MOS 68G10)
Occupational Therapy Specialist (Five)—(SPC/E4, MOS 68W10/N3)
Mental Health Specialist (Ten)—(Six SPC/E4 and Four SPC/E3, MOS 68X10)

SECTION II — MEDICAL DETACHMENT, COMBAT STRESS CONTROL (TOE 08463L000) (MEDICAL FORCE 2000)

MISSION

C-11. The mission of the AOE MF2K medical detachment, CSC is to provide COSC interventions and activities to supported units in its AO. It augments division and brigade MH sections; provides direct support to combat brigades without organic BH officers; and provides area support in its AO. The medical detachment, CSC reconstitutes other brigade and division COSC assets. The medical detachment, CSC provides COSC interventions and activities to indigenous populations as directed in stability and reconstruction operations, to include domestic support operations, humanitarian assistance, disaster relief, peace support operations, and detention facility operations. The medical detachment, CSC provides COSC interventions and activities between deployments to units in support of their readiness preparation and throughout their deployment cycle. Under MRI this detachment will be replaced with medical detachment, CSC, TOE 08463A000.

> ***Note.*** The MF2K and MRI medical detachments, CSC conduct COSC support operations very similarly. However, in the MRI medical detachment, CSC, psychiatry assets are assigned to the restoration section and psychology assets are assigned to the preventive section whereas in the MF2K detachment, the manning strategy for psychology and psychiatry is reversed: psychologists are in the restoration section and psychiatrists are on the preventive teams.

ASSIGNMENT

C-12. The MF2K medical detachment, CSC (TOE 08463L000), is assigned to a corps MEDBDE (TOE 08422A100), or a corps MEDCOM (TOE 08411A000), or other medical task force C2 elements. Its teams may be attached to a medical company, CSC (TOE 08467L000); an ASMB (TOE 08456A000); a brigade or divisional medical company; or a CSH.

EMPLOYMENT

C-13. Detachment headquarters is usually located in the DSA. Its teams disperse and are employed throughout its AO.

CAPABILITIES

C-14. The MF2K medical detachment, CSC provides—

- Mobile COSC interventions from the preventive section which can divide into three preventive teams.
- Holding for 40 Soldiers by the restoration team for COSC interventions such as restoration. With additional logistical support, the team can accommodate additional Soldiers under surge conditions.

STAFF RESPONSIBILITIES

C-15. The medical detachment assists the C2 headquarters (to which it is assigned or attached) regarding planning and coordination of COSC support, stress threat, mental and physical stressors, stress behaviors, principles of COSC, and implementation of COSC functional areas.

DEPENDENCY

C-16. This unit is dependent on—

- Appropriate elements of the supporting unit for FHP; religious support; legal; finance; field feeding; personnel and administrative services support; laundry and clothing exchange; mortuary affairs support; and security of EPW, detainee, and US prisoner patients.
- Supporting unit for medical administration; logistics (including MEDLOG); medical regulating; evacuation; coordination for RTD; and unit-level equipment and CE maintenance.

MOBILITY AND SECURITY

C-17. This unit is 100-percent mobile. It requires 100 percent of its organic personnel and equipment be transported in a single lift, using its organic vehicles. Upon relocation, Soldiers being held will require additional transportation. This unit is responsible for perimeter defense of its immediate operational area. However, it is dependent on appropriate elements of the supporting unit for additional security, to include security of convoy operations. Personnel of the detachment are provided weapons for their personal defense and for the defense of their patients and/or held Soldiers.

ORGANIZATION

C-18. This 25-person unit (see Table C-4) is composed of a headquarters, a combat stress preventive section with three preventive teams, and a CSCR. The modular CSC teams found in the MF2K CSC medical detachment are similar to those found in the CSC medical company.

Table C-4. Detachment teams

Headquarters	*Preventive Section*	*Restoration Section*
	CSCP#1	CSCR
	CSCP#2	
	CSCP#3	

C-19. The detachment headquarters provides C2 for the detachment. It is responsible for planning, coordinating, and implementing COSC interventions and activities for supported units. It consists of three personnel: a detachment Commander (05, 60W00, MC), a detachment sergeant (E7, 68X40), and a wheeled-vehicle mechanic (E4, 63B10). The detachment commander also serves as a treating physician with the preventive section. The detachment NCOIC (a senior MH NCO) also serves as the restoration team sergeant. Detachment officers and NCOs from the prevention team and the restoration team may be assigned additional duties, which enhance the overall effectiveness of the headquarters section. See Table C-5 for personnel assigned to detachment headquarters.

Table C-5 Detachment headquarters personnel

Commander (LTC/05, AOC 60W00)
Detachment Sergeant (SFC/E7, MOS 68X40)
Wheeled Vehicle Mechanic (SPC/E4, MOS 63B10)
Note. The commander also serves as the psychiatrist in the preventive section.

Preventive Section

C-20. The 12-person preventive section staff is identified in Table C-6. The preventive section may be task-organized to conduct its COSC mission. Preventive section personnel may also be task-organized with personnel of the restoration section into teams for specific missions. The preventive section can divide into three preventive teams. The section (and team) leader position may be held by any of the officers assigned to the section. This section's COSC interventions and activities are—

- Unit needs assessment; consultation and education; critical event and transition management; COSC triage; stabilization (emergency); and BH treatment.
- Assisting with restoration and reconditioning at the CSC detachment program. Overseeing a 1- to 3-day COSC restoration program in a brigade, division, or ASMC holding section or in another area suitable for Soldiers experiencing COSR and/or other stress-related disorders.

Table C-6. Preventive section 7/0/6

Psychiatrist (LTC/05, AOC 60W00)
Clinical Psychologist (Three)—(MAJ/04, AOC 73B67)
Social Worker (Three)—(CPT/03, AOC 73A67)
Team Chief (Three)—(SGT/E5, MOS 68X20)
Mental Health Specialist (Three)—(SPC/E4, MOS 68X10)
Note. The psychiatrist is also counted in the headquarters section.

Restoration Team

C-21. The 9-person restoration team staff is identified in Table C-7. The restoration team is task-organized to provide COSC interventions and activities. Restoration team personnel may also be task-organized with personnel of the CSC preventive section into teams for specific missions. Each restoration team can deploy a four-person mobile team with a HMMWV. The CSCF usually collocates with a supported divisional medical company to provide mobile COSC support within a DSA and conduct restoration programs, as required. The CSCF provides staff and equipment for operating a restoration or reconditioning center. This section's COSC interventions and activities are—

- Unit needs assessment; consultation and education; critical event and transition management; COSC triage; stabilization; and BH treatment.
- Conducting Soldier restoration and reconditioning programs.
- Assisting the CSH psychiatric section when psychiatric ward capability is required (refer to FM 8-10-14).

Table C-7. Restoration team 3/0/6

Psychiatrist (MAJ/04, AOC 60W00)
Psychiatric/Mental Health Nurse (MAJ/04, AOC 66COO/7T)
Occupational Therapist (CPT/03, AOC 65A00)
Occupational Therapy NCO (SSG/E6, 68W30/N3)
Team Chief (SSG/E6, MOS 68X30)
Occupational Therapy Sergeant (SGT/E5, MOS 68W20/N3)
Mental Health NCO (SGT/E5, MOS 68X20)
Mental Health Specialist (Two)—(One SPC/E4 and One PFC/E3, MOS 68X10)

Glossary

SECTION I — ACRONYMS AND ABBREVIATIONS

AAR	after-action review
ABCA	American, British, Canadian, and Australian
ACR	armored cavalry regiment
AMEDD	Army Medical Department
AN	Army Nurse Corps
AO	area of operations
AOC	area of concentration
AOE	Army of Excellence
AR	Army regulation
ART	Army tactical task
ASMB	area support medical battalion
ASMC	area support medical company
AUTL	Army Universal Task List
BAS	battalion aid station
BCT	brigade combat team
BDP	behavioral disordered patient
BF	battle fatigue
BH	behavioral health
BICEPS	brevity, immediacy, contact, expectancy, proximity, and simplicity (See Section II for definition.)
BSA	brigade support area
BSB	brigade support battalion
BSMC	brigade support medical company
C	Centigrade
C2	command and control
C4I	command, control, communications, computers, and intelligence
CBRN	chemical, biological, radiological, and nuclear
CE	communications-electronics
CH	chaplain
CHL	combat health logistics
CHS	combat health support
COA	course(s) of action
CONUS	continental United States
COSC	combat and operational stress control (See Section II for definition.)
COSR	combat and operational stress reaction (See Section II for definition.)

CP	command post
CPT	captain
CRST	combat religious support team
CS	combat support
CSC	combat stress control
CSCF	combat stress control fitness
CSCP	combat stress control preventive
CSCR	combat stress control restoration
CSH	combat support hospital
CSR	combat stress reaction
CZ	combat zone
DA	Department of the Army
DD	Department of Defense
DEPMEDS	Deployable Medical Systems
DISCOM	division support command
DMHS	division mental health section
DNBI	disease and nonbattle injury(ies)
DOD	Department of Defense
DODD	Department of Defense Directive
DODI	Department of Defense Instruction
DS	direct support
DSA	division support area
EAB	echelons above brigade
EAC	echelons above corps
EAD	echelons above division
EMT	emergency medical treatment
EPW	enemy prisoner(s) of war
ETOD	end-of-tour debriefing
F	Fahrenheit
FAP	Family Advocacy Program
FHP	force health protection
1LT	first lieutenant
1SG	first sergeant
FM	field manual
FMC	United States Field Medical Card (Department of Defense Form 1380)
FRAGO	fragmentary order
FSB	forward support battalion
FSMC	forward support medical company
G-1	Assistant Chief of Staff (Personnel)
G-2	Assistant Chief of Staff (Intelligence)
G-3	Assistant Chief of Staff (Operations and Plans)

G-4	Assistant Chief of Staff (Logistics)
G-5	Assistant Chief of Staff (Plans and Policy)
G-9	Assistant Chief of Staff (Civil-Military Affairs)
GP	general purpose
HBCT	heavy brigade combat team
HHD	headquarters and headquarters detachment
HIP	help-in-place
HMMWV	high-mobility multipurpose wheeled vehicle
HSL	health service logistics
HSS	health service support
IED	improvised explosive device
JAG	Judge Advocate General
LTC	lieutenant colonel
MAJ	major
MC	Medical Corps
MCW	minimal care ward
MEDBDE	medical brigade
MEDCEN	medical center
MEDCOM	medical command (US Army)
MEDDAC	medical department activity
MEDLOG	medical logistics
METT-TC	mission, enemy, terrain and weather, troops and support available, time available, and civil considerations
MF2K	Medical Force 2000
mg	milligram(s)
MH	mental health
MHAT	Mental Health Advisory Team (US Army)
MMB	multifunctional medical battalion
MOS	military occupational specialty
MP	military police
MRI	Medical Reengineering Initiative
MRO	medical regulating officer
MS	Medical Service Corps
MSB	main support battalion
MSMC	main support medical company
MTF	medical treatment facility
MTOE	modification table of organization and equipment
NATO	North Atlantic Treaty Organization
NCO	noncommissioned officer
NCOIC	noncommissioned officer in charge
NP	neuropsychiatric

OEF	Operation Enduring Freedom
OIF	Operation Iraqi Freedom
OPCON	operational control
OPLAN	operation plan
OPORD	operation order
OPTEMPO	operational tempo
OSAT	operational stress assessment team
OSR	operational stress reaction
OT	occupational therapy
PA	physician assistant
PCP	phencyclidine hydrochloride
PD	psychological debriefing
PERSTEMPO	personnel tempo
PIES	proximity, immediacy, expectancy, and simplicity
POC	point of contact
PTE	potentially traumatizing event
PTSD	posttraumatic stress disorder
pub	publication
PVNTMED	preventive medicine
R&R	rest and relaxation
RC	Reserve Component
RTD	return to duty
S1	Adjutant (US Army)
S2	Intelligence Officer (US Army)
S3	Operations and Training Officer (US Army)
S4	Supply Officer (US Army)
SBCT	Stryker Brigade Combat Team
SGT	sergeant
SOI	signal operation instructions
SOP	standing operating procedure
SP	Army Medical Specialty Corps
SPC	specialist
SRC	standard requirement code
SSG	staff sergeant
TC	training circular
TDA	table of distribution and allowances
TEM	traumatic events management
TEMPER	tent, expandable, modular, personnel
TG	technical guide
TOE	table(s) of organization and equipment
TRADOC	United States Army Training and Doctrine Command

TSOP	tactical standing operating procedures
UCMJ	Uniform Code of Military Justice
UMT	unit ministry team
UNA	unit needs assessment
US	United States
USAMEDDC&S	United States Army Medical Department Center and School
USAMRMC	United States Army Medical Research and Materiel Command
USAR	United States Army Reserve
USMC	United States Marine Corps
VHA	Veterans Health Adminstration

SECTION II — TERMS

brevity, immediacy, contact, expectancy, proximity, and simplicity

An acronym used for the management of combat and operational stress reactions—*brevity* (usually less than 72 hours); *immediacy* (as soon as symptoms are evident); *contact* (chain of command remains directly involved in the Soldiers recovery and return to duty), *expectancy* (combat stress control unit personnel expectation that casualties will recover); *proximity* (of treatment at or as near the front as possible); *simplicity* (the use of simple measures such as rest, food, hygiene, and reassurance). Also known as BICEPS.

combat and operational stress control

A coordinated program for the prevention of and actions taken by military leadership to prevent, identify, and manage adverse combat and operational stress reactions in units. Also known as COSC.

combat and operational stress reaction

The expected, predictable, emotional, intellectual, physical, and/or behavioral reactions of Service members who have been exposed to stressful events in combat or military operations other than war. Also known as COSR.

5 R's

Actions used for combat and operational stress reaction control that include—**R**eassure of normality; **R**est (respite from combat or break from the work); **R**eplenish bodily needs (such as thermal comfort, water, food, hygiene, and sleep); **R**estore confidence with purposeful activities and contact with his unit; **R**eturn to duty and reunite Soldier with his unit.

reconditioning program

An intensive 4- to 7-day program (may be extended by exception to theater evacuation policy) of replenishment, physical activity, therapy, and military retraining for combat and operational stress control casualties and neuropsychiatric cases (including alcohol and drug abuse) who require successful completion for return to duty or is evacuated for further neuropsychiatric evaluation.

Soldier restoration

A 24- to 72-hour (1- to 3-day) program in which Soldiers with combat and operational stress reactions receive treatment.

stabilization

The initial short-term management and evaluation of severely behaviorally disturbed Soldiers caused by an underlying combat and operational stress reaction, behavioral health disorder, or alcohol and/or drug abuse reaction.

References

SOURCES USED

These are the sources quoted or paraphrased in this publication.

DEPARTMENT OF DEFENSE

These documents are available online at: http://www.dtic.mil/whs/directives/

DODD 6490.1, *Mental Health Evaluations of Members of the Armed Forces*, 1 October 1997

DODD 6490.2, *Comprehensive Health Surveillance*, 21 October 2004

DODD 6490.5, *Combat Stress Control (CSC) Program*, 23 February 1999

DODI 6490.4, *Requirements for Mental Health Evaluations of Members of the Armed Forces,* 28 August 1997

MULTISERVICE PUBLICATION

FM 6-22.5/MCRP 6-11C/NTTP 1-15M, *Combat Stress*, 23 June 2000

ARMY PUBLICATIONS

These publications are available online at: https://akocomm.us.army.mil/usapa

AR 40-216, *Neuropsychiatry and Mental Health,* 10 August 1984

FM 7-15, *The Army Universal Task List*, 31 August 2003

FM 22-51, *Leaders' Manual for Combat Stress Control,* 29 September 1994

DOCUMENTS NEEDED

These documents must be available to the intended users of this publication.

DEPARTMENT OF DEFENSE FORMS

These forms are available online at: https://akocomm.us.army.mil/usapa

DD Form 1380, *US Field Medical Card*

DD Form 2795, *Pre-Deployment Health Assessment*

DD Form 2796, *Post-Deployment Health Assessment*

MULTISERVICE PUBLICATION

AR 190-8/OPNAVINST 3461.6/AFJI 31-304/MCO 3461.1, *Enemy Prisoners of War, Retained Personnel, Civilian Internees, and Other Detainees,* 1 October 1997

ARMY PUBLICATIONS

These publications are available online at: https://akocomm.us.army.mil/usapa

AR 25-1, *Army Knowledge Management and Information Technology Management*, 15 July 2005

AR 40-3, *Medical, Dental, and Veterinary Care,* 12 November 2002

AR 40-61, *Medical Logistics Policies*, 28 January 2005

AR 40-66, *Medical Record Administration and Health Care Documentation,* 20 July 2004

AR 40-68, *Clinical Quality Management,* 26 February 2004

AR 40-400, *Patient Administration*, 12 March 2001

AR 40-501, *Standards of Medical Fitness,* 1 February 2005

AR 340-21, *The Army Privacy Program*, 5 July 1985

DA Pamphlet 611-21, *Military Occupational Classification and Structure*, 31 March 1999

FM 4-02, *Force Health Protection in a Global Environment,* 13 February 2003

FM 4-02.6, *The Medical Company—Tactics, Techniques, and Procedures,* 1 August 2002 (Change 1, 9 April 2004)

FM 4-02.7, *Health Service Support in a Nuclear, Biological, and Chemical Environment—Tactics, Techniques, and Procedures,* 1 October 2002

FM 4-02.10, *Theater Hospitalization*, 3 January 2005

FM 4-02.16, *Army Medical Information Management—Tactics, Techniques, and Procedures,* 22 August 2003

FM 4-02.17, *Preventive Medicine Services,* 28 August 2000

FM 4-25.12, *Unit Field Sanitation Team,* 25 January 2002

FM 8-10-6, *Medical Evacuation in a Theater of Operations—Tactics, Techniques, and Procedures*, 14 April 2000

READINGS RECOMMENDED

These sources contain relevant supplemental information.

JOINT OR MULTISERVICE PUBLICATIONS

Most joint publications are available online at: http://www.dtic.mil/doctrine/

Joint Pub (JP) 1-02, *Department of Defense Dictionary of Military and Associated Terms,* 12 April 2001 (As amended through 9 June 2004)

JP 3-0, *Doctrine for Joint Operations*, 9 September 2001

JP 4-02, *Doctrine for Health Service Support in Joint Operations*, 30 July 2001

FM 1-02/MCRP 5-12A, *Operational Terms and Graphic*, 21 September 2004

FM 4-25.11/NTRP 4-02.1.1/AFMAN 44-163(I)/MCRP 3-02G), *First Aid,* 23 December 2002 (Change 1, 15 July 2004)

FM 8-9/NAVMED P-5059/AFJMAN 44-151V1V2V3, *NATO Handbook on the Medical Aspects of NBC Defensive Operations AMedP-6(B), Part I—Nuclear, Part II—Biological, Part IIII—Chemical,* 1 February 1996

FM 8-284/NTRP 4-02.23(NAVMED P-5042)/AFMAN (I) 44-156/MCRP 4-11.1C, *Treatment of Biological Warfare Agent Casualties,* 17 July 2000 (Change 1, 8 July 2002)

FM 8-285(4-02.285)/NAVMED P-5041/AFJMAN 44-149/FMFM 11-11, *Treatment of Chemical Agent Casualties and Conventional Military Chemical Injuries,* 22 December 1995

FM 21-10/MCRP 4-11.1D, *Field Hygiene and Sanitation,* 21 June 2000

ARMY PUBLICATIONS

Most of these publications are available online at: https://akocomm.us.army.mil/usapa

AR 71-32, *Force Development and Documentation—Consolidated Policies*, 3 March 1997

AR 635-200, *Active Duty Enlisted Administrative Separations*, 6 June 2005

FM 3-0, *Operations*, 14 June 2001

FM 3-06.11, *Combined Arms Operations in Urban Terrain*, 28 February 2002

FM 3-07, *Stability Operations and Support Operations*, 20 February 2003

FM 3-19.40, *Military Police Internment/Resettlement Operations*, 1 August 2001

FM 3-21.21, *The Stryker Brigade Combat Team Infantry Battalion*, 8 April 2003 (Change 1, 31 July 2003)

FM 3-21.31, *The Stryker Brigade Combat Team*, 13 March 2003

FM 3-90.3, *The Mounted Brigade Combat Team*, 1 November 2001

FM 5-0, *Army Planning and Orders Production*, 20 January 2005

FM 4-0, *Combat Service Support*, 29 August 2003

FM 4-02.4, *Medical Platoon Leaders' Handbook—Tactics, Techniques, and Procedures*, 24 August 2001 (Change 1, 18 December 2003)

FM 4-02.19, *Dental Service Support in the Theater of Operations*, 1 March 2001

FM 4-02.21, *Division and Brigade Surgeons' Handbook (Digitized)—Tactics, Techniques, and Procedures*, 15 November 2000

FM 4-02.24, *Area Support Medical Battalion— Tactics, Techniques, and Procedures*, 28 August 2000

FM 4.25.10, *Field Hygiene and Sanitation*, 20 June 2002

FM 8-10-14, *Employment of the Combat Support Hospital Tactics, Techniques, and Procedures*, 29 December 1994 (will be revised as FM 4-02.14)

FM 8-42, *Combat Health Support in Stability Operations and Support Operations*, 27 October 1997 (will be revised as FM 4-02.42)

FM 8-50, *Prevention and Medical Management of Laser Injuries*, 8 August 1990 (will be revised as FM 4-02.50)

FM 8-55, *Planning for Health Service Support,* 9 September 1994 (will be revised as FM 4-02.55)

FM 10-27-4, *Organizational Supply and Services for Unit Leaders,* 14 April 2000 (will be revised as FM 4-20.05)

FM 27-1, *Legal Guide for Commanders.* 13 January 1992 (will be revised as FM 1-04.1)

FM 27-10, *The Law of Land Warfare,* 18 July 1956 (Reprinted with basic including Change 1, 15 July 1976) (will be revised as FM 1-04.10)

FM 63-20, *Forward Support Battalion*, 26 February 1990 (will be revised as FM 4-93.20)

FM 63-21, *Main Support Battalion*, 7 August 1990 (will be revised as FM 4-93.21)

FM 100-9, *Reconstitution*, 13 January 1992 (will be revised as FM 4-100.9)

FM 100-17, *Mobilization, Deployment, Redeployment, and Demobilization,* 28 October 1992 (will be revised as FM 3-35)

FM 100-17-5, *Redeployment*, 29 September 1999 (will be revised as FM 3-35)

MEDCOM Regulation 40-38, *Command-Directed Mental Health Evaluation*, 1 September 2001

TC 25-20, *A Leader's Guide to After-Action Reviews,* 30 September 1993

TRADOC Pamphlet 600-22, *Leaders Guide for Suicide Prevention Planning,* 16 February 2005 (Available at http://www.tradoc.army.mil/publications.htm)

United States Army Center for Health Promotion and Preventive Medicine

Most of these publications are available online at: http://chppm-www.apgea.army.mil/dhpw/

USACHPPM TG 240, *Combat Stress Behaviors,* June 2004

USACHPPM TG 241, *Combat Operational Stress Reaction* (*COSR*) (*"Battle Fatigue"*), June 2004

USACHPPM TG 242, *Combat Operational Stress Reactions* (*COSR*) *Prevention: Leader Actions,* June 2004

USACHPPM TG 243, *Combat Stress Card*, May 1999

Tables of Organization and Equipment

TOE 08411A000, *Medical Command, Corps* (*MRI*)

TOE 8422A100, *Medical Brigade, Corps* (*MRI*)

TOE 08456A000, *Area Support Medical Battalion*

TOE 08457A000, *Area Support Medical Company*

TOE 08463A000, *Medical Detachment, Combat Stress Control* (*MRI*)

TOE 08463L000, *Medical Detachment, Combat Stress Control* (*MF2K*)

TOE 08467L000, *Medical Company, Combat Stress Control* (*MF2K*)

TOE 08660G000, *Medical Detachment, Combat Operational Stress Control* (*Force Development Update*)

DEPARTMENT OF VETERANS AFFAIRS

Veterans Health Administration, *Clinical Practice Guidelines (Mental Health),* available online at: http://www.oqp.med.va.gov/cpg

Index

References are to paragraph numbers except where otherwise specified.

PREFACE

Combat Stress is the mental, emotional or physical tension, strain, or distress resulting from exposure to combat and combat-related conditions. Controlling combat stress is a command responsibility. In terms of Service members lost from action and reduced performance, combat stress seriously affects mission accomplishment. It is a leader's responsibility to take action to strengthen Service members' tolerance to combat stress and manage it in his or her unit.

Combat stress reactions are the result of exposure to the same conditions during military actions that cause physical injury and disease in battle or its immediate aftermath, and many combat stress reactions occur in persons who are also wounded or ill with disease. Rates of combat stress casualties vary greatly, with higher ratios during lengthy periods of intense combat. In Okinawa 1945, during a peak month of battle, the combat stress casualties among Marine Forces were reported as high as one for every two wounded in action (WIA). Under less lengthy periods, as suggested by data acquired from the Israeli Defense Forces fighting in Lebanon 1982, the ratio of combat stress casualties to WIA in small units can be as high as one to one. In the past, we have generally suffered as many as one battle stress casualty for every three to five WIA in heavy fighting. However, highly trained units with strong leadership and high esprit de corps have fewer combat stress casualties.

While this manual focuses on combat-induced stress reactions, it is important to emphasize that "combat stress" is not restricted only to combat, but may also arise from combat-like conditions present during military operations other than war. In an area of operations characterized by continuous action and high danger, our forces may experience high rates of stress casualties unless small-unit leaders are trained and prepared to manage stress.

This publication is written to inform small-unit leaders of stress characteristics and management techniques in order to *prevent, reduce, identify,* and *treat* combat stress reactions in the Service member's own unit to the maximum extent possible. A significant part of training is learning to control and cope with stress. Leaders must learn to cope with their own stress and then assist junior personnel in managing their stress. The application of combat stress management techniques helps conserve fighting strength and provides one more step toward achieving success.

Table of Contents

Chapter 2. Combat Stress Prevention, Management, and Control

Chapter 3. Command Leadership Actions

Chapter 4. Sleep Deprivation

Appendices

Chapter 1

Combat Stress Identification

1001. INTRODUCTION

The Marine Corps' success as a fighting force is dependent on leadership that maintains a balanced focus between mission accomplishment and troop welfare. The small-unit leader is the key to building and maintaining high unit morale and peak efficiency. He achieves this in part by knowing his troops and understanding their strengths and weaknesses. To maintain that same level of morale and efficiency in combat, the small-unit leader must understand how to recognize, prevent, and even personally contend with reactions to combat stress when it occurs in his unit. If a condition accounted for as many casualties in combat and the condition was at least partially preventable, the prudent combat leader would be interested in knowing more about it. Combat stress reaction(s), also called battle fatigue, is that condition. It has the potential to disable the most courageous Service member and influence the success or failure of a unit in accomplishing its mission.

1002. HISTORY

During the 1942-45 period in the European Theater, there was a ratio of one combat stress casualty for every three WIA. In a month of especially horrible, continuous fighting in Okinawa in 1945, the 6th Marine division had one stress casualty for every 1.8 WIA. However, the airborne divisions in Europe never had more than one for ten WIA, and usually less, even though they experienced very high casualties in some battles.

To accomplish a mission successfully, planners must use some guidelines to estimate losses from combat stress. What are the reasons for the tremendous range between high and low battle stress casualty rates? While the answer to this question is complex, it is clear that better-trained troops have fewer killed and wounded, and proportionally fewer stress casualties. Service members—especially leaders—can learn to recognize the symptoms and prevent or reduce the disruptive effects of combat stress.

1003. REACTIONS TO COMBAT STRESS

Service members exposed to danger experience physical and emotional reactions that are not present under more tranquil circumstances. Some reactions sharpen abilities to survive and win; other reactions may produce disruptive behaviors and threaten individual and unit safety. These adverse behaviors are collectively called *combat stress reaction*. The operative word is "behaviors." People in combat experience a range of emotions, but their behavior influences immediate safety and mission success.

Combat and combat-related military missions can also impose combinations of heavy physical work; sleep loss; dehydration; poor nutrition; severe noise, vibration, and blast; exposure to heat, cold or wetness; poor hygiene facilities; and perhaps exposure to infectious diseases, toxic fumes or substances. These, in combination with other influences—such as concerns about problems back home—affect the ability to cope with the perception of danger, and diminish the skills needed to accomplish the mission. Environmental stressors often play an important part in causing the adverse or disruptive combat stress reaction behaviors. The leader must work to keep each Service member's perception of danger balanced by the sense that the unit has the means to prevail over it. The leader must keep himself and his unit working at the level of stress that sustains performance and confidence. When troops begin to lose confidence in themselves and their leader, adverse stress reactions are most likely to occur.

It is important for the small-unit leader to recognize these adverse behaviors at the onset in order to intervene promptly for the safety and benefit of individual Service members and the unit. These behaviors may take many forms and can range from subtle to dramatic. Any Service member who shows persistent, progressive behavior that deviates from his baseline behavior may be demonstrating the early warning signs and symptoms of a combat stress reaction. Trying to memorize every possible sign and symptom is less useful to prompt diagnosis than to keep one simple rule in mind: *Know your troops, and be alert for any sudden, persistent or progressive change in their behavior that threatens the functioning and safety of your unit.*

1004. OBSERVING AND RECOGNIZING COMMON REACTIONS TO COMBAT STRESS

Ranges of fatigue, fear, anxiety, and depression affect most Service members in combat and in some military operations other than war. Mild stress reaction may be signaled by changes in behavior and only be discernible by the person himself or by close comrades. The unit leader and medical personnel depend on information from the Service member or his comrades for early recognition of combat stress reactions to provide prompt and appropriate help. Table 1-1 lists some mild stress reactions.

Table 1-1. Mild Stress Reactions.

PHYSICAL	EMOTIONAL
Trembling	Anxiety, indecisiveness
Jumpiness	Irritability, complaining
Cold sweats, dry mouth	Forgetfulness, inability to concentrate
Insomnia	Nightmares
Pounding heart	Easily startled by noise, movement, and light
Dizziness	Tears, crying
Nausea, vomiting or diarrhea	Anger, loss of confidence in self and unit
Fatigue	
"Thousand-yard" stare	
Difficulty thinking, speaking, and communicating	

Severe stress reactions may prevent the individual from performing his duties or create a concern for personal safety or the safety of others. A variety of more serious reactions or warning signs are listed in Table 1-2. These do not necessarily mean that the person must be relieved from duty, but warrant immediate evaluation and help by the small-unit leader.

Table 1-2. Severe Stress Reactions.

PHYSICAL	EMOTIONAL
Constantly moves around	Talks rapidly and/or inappropriately
Flinches or ducks at sudden sound and movement	Argumentative; acts recklessly
Shakes, trembles	Indifferent to danger
Cannot use part of body (hand, arm leg) for no apparent physical reason	Memory loss
Inability to see, hear or feel	Stutters severely, mumbles or cannot speak at all
Is physically exhausted; cries	Insomnia; severe nightmares
Freezes under fire or it totally immobile	Sees or hears things that do not exist
States vacantly, staggers or sways when standing	Has rapid emotional shifts
Panics, runs under fire	Socially withdrawn
	Apathetic
	Hysterical outbursts
	Frantic or strange behavior

Fatigue

The most common stress reactions include:

- Slow reaction time.
- Difficulty sorting out priorities.
- Difficulty starting routine tasks.
- Excessive concern with seemingly minor issues.
- Indecision, difficulty-focusing attention.
- A tendency to do familiar tasks and preoccupation with familiar details. These reactions may reach a point where the person becomes very passive, or wanders aimlessly.
- Loss of initiative with fatigue and exhaustion.

Muscular Tension

Headaches, Backache

Muscular tension increases strain on the scalp and spine and often leads to headaches, pain, and cramps.

Inability to Relax

Prolonged muscular tension wastes energy and leads to fatigue and exhaustion. Muscles must relax periodically to enable free blood flow, waste product flushing, and nutrient replenishment.

Shaking and Tremors

Mild Shaking

During incoming rounds, the individual may experience mild shaking. This symptom appears and disappears rapidly and is considered a normal reaction to conditions of great danger.

Marked or Violent Shaking

A common post-battle reaction, marked or violent shaking can be incapacitating if it occurs during the action. If shaking persists long after the precipitating stimulus ceases, or if there was no stimulus, the individual should be checked by medical personnel.

Perspiration

It is normal to experience either mild or heavy sweating or sensations of chilliness under combat stress.

Digestive and Urinary Systems

Nausea and Vomiting

"Butterflies in the stomach" is a common stress feeling. Vomiting may occur as a result of an extreme experience like that of a fire-fight, shelling, etc., or in anticipation of danger.

Loss of Appetite

Appetite loss may result as a reaction to stress. It becomes a significant problem if rapid weight loss occurs, or the person does not eat a sufficiently balanced diet to keep his muscles and brain supplied for sustained operations.

Abdominal Distress

Acute abdominal pain ("knotted stomach," "heartburn") may occur during combat. Persistent and severe abdominal pain is a disruptive reaction, and may indicate a medical condition.

Frequent Urination

Frequent urination may occur, especially at night.

Incontinence

During extremely dangerous moments, inability to control bowel and/or bladder functions may occur. Incontinence is embarrassing, but it is not abnormal under these circumstances.

Circulatory and Respiratory Systems

Heart Palpitations

Rapid heartbeat, a sense of pressure in the chest, occasional skipped beats, and sometimes chest pains are common with anxiety or fear. Very irregular heartbeats need to be checked by medical personnel.

Hyperventilation

Hyperventilation is identified by rapid respiration, shortness of breath, dizziness, and a sense of choking. It is often accompanied with tingling and cramping of fingers and toes. Simple solutions are increased exercise and breathing with a paper bag over the nose and mouth; or breathing slowly using abdominal muscles (called "abdominal breathing").

Faintness and Giddiness

These reactions occur in tandem with generalized muscular weakness, lack of energy, physical fatigue, and extreme stress. Brief rest should be arranged, if possible.

Sleep Disturbance

Difficulty Falling Asleep

Sometimes a Service member who has experienced intense battle conditions often cannot fall asleep even when the situation permits, or, when he does fall asleep, he frequently wakes up and has difficulty getting back to sleep.

Nightmares

Terror dreams, battle dreams, and nightmares of other kinds cause difficulty in staying asleep. Sleep disturbances in the form of dreams are part of the coping process. This process of *working through* combat experiences is a means of increasing the level of tolerance of combat stress. The individual may have battle-related nightmares or dreams that a close relative (spouse, parent) or another person important in his life has been killed in the battle. As time passes, the nightmares usually occur with less intensity and less frequency. In some cases, a Service member, even when awake, may re-experience the memory of the stressful experience as if it were recurring (called a "flashback"). This is usually triggered by a smell, sound or sight, and is not harmful as long as the Service member realizes it is only a memory and does not react inappropriately or feel overwhelmed. However, if it happens frequently or is very distressing, help should be sought from the chaplain or medical.

Restless Sleep

When a person is asleep, the sleep is not restful sleep if the person is constantly being half-wakened by noise, movement, or other stimuli. Heavy snoring often indicates poor quality sleep. The individual wakes up as tired as when he went to sleep. Finding a more comfortable position, away from distractions, can help.

Excessive Sleep

Individuals exhibiting a need for excessive sleep may be exhibiting symptoms of combat stress; however, excessive sleep is also a sign of substance abuse or depression (Persistent insomnia is a more common indicator of possible depression.)

Visual/Hearing Problems and Partial Paralysis

Stress-related blindness, deafness, loss of other sensations, and partial paralysis are not true physical injuries, but physical symptoms that unconsciously enable the individual to escape or avoid a seemingly intolerably stressful situation. These symptoms can quickly improve with reassurance and encouragement from comrades, unit medical personnel or battalion physician. If they persist, the physician must examine the individual to be sure there is not a physical cause; for example, laser range finders can cause temporary or partial blindness, and nearby explosions can cause ear damage. Individuals with these physical conditions are unaware of the causative relationship with their inability to cope with stress. They are honestly concerned with their physical symptoms and want to get better. They are willing to discuss them, and do not mind being examined. This is contrary to "malingerers" faking a physical illness, who are often reluctant to talk, or over-dramatize their disability and refuse an examination.

Visual problems include blurred vision, double vision, difficulty in focusing, or total blindness. Hearing problems include inability to hear orders and/or nearby conversations or complete deafness occurs. Paralysis or loss of sensation is usually confined to one arm or leg. Prickling sensations or rigidity of the larger joints occur. However, temporary complete immobility (with normal breathing and reflexes) can occur. If these reactions do not recover quickly with immediate reassurance, care must be taken in moving the casualty to medical for an evaluation to avoid making a possible nerve or spinal cord injury worse.

Bodily Arousal

Not all emotional reactions to stress are necessarily negative. For example, the body may become aroused to a higher degree of awareness and sensitivity.

Threat

In response to threat, the brain sends out chemicals arousing the various body systems. The body is ready to fight or take flight. The alerting systems of the experienced combat veteran become finely tuned, so that he may ignore loud stimuli that pose no danger (such as the firing of nearby friendly artillery). However, he may awaken from sleep at the sound of an enemy mortar being fired and take cover before the round hits. The senses of vision and smell can also become very sensitive to warning stimuli. The Service member may instantly focus and be ready to react.

Hyperalert

Hyperalert refers to being distracted by any external stimuli that might signal danger, and overreacting to things that are, in fact, safe. The hyperalert Service member is not truly in tune with his environment, but is "on a hair trigger." He is likely to over-react, misinterpret reassuring information as threats, and react without adequate critical thinking. Consequences can range from firing at an innocent noise to designating as hostile an innocent target.

Startle Reactions

Startle reactions are part of an increased sensitivity to minor external stimuli (on-guard reactions). Leaping, jumping, cringing, jerking, or other forms of involuntary self-protective motor responses to sudden noises are noted. The noises are not necessarily very loud. Sudden noise, movement, and light cause startle reactions; for example, unexpected movement of an animal (or person) precipitates weapon firing.

Anxiety (Fear of Death, Pain, and Injury)

Fear of death, pain, and injury causes anxiety reactions. After witnessing the loss of a comrade in combat, a Service member may lose self-confidence and feel overly vulnerable or incapable. The death of a buddy leads to serious loss of emotional support. Feelings of "survivor guilt" are common. The survivors each brood silently, second-guessing what they think they might have done differently to prevent the loss. While the Service member feels glad he survived, he also feels guilty about having such feelings. Understanding support and open grieving shared within the unit can help alleviate this.

Irritability

Reaction

Mild irritable reactions range from angry looks to a few sharp words, but can progress to more serious acts of violence. Mild irritability is exhibited by sharp, verbal overreaction to normal, everyday comments or incidents; flare-ups involving profanity; and crying in response to relatively slight frustrations.

Explosion of Aggressive Behavior

Sporadic and unpredictable explosions of aggressive behavior (violence) can occur with little or no provocation. For example, a

Service member tries to pick a fight with another Service member. The provocation may be a noise, such as the closing of a window, an accidental bumping or normal, verbal interaction.

Short Attention Span

Persons under stress have short attention spans. They find it difficult to concentrate. Short attention span causes a Service member to have difficulty following orders. The Service member does not easily understand what others are saying. The person has difficulty following directions, aiding others or performing unfamiliar tasks.

Depression

People respond to stress with protective defensive reactions against painful perceptions. Emotional dulling or numbing of normal responsiveness is a result. The reactions are easily observed changes from the individual's usual self.

Low Energy Level

The observer notes decreased effectiveness on the job, decreased ability to think clearly, excessive sleeping or difficulty falling asleep, and chronic tiredness. Emotions such as pride, shame, hope, grief, and gratitude no longer matter to the person.

Social Withdrawal

A Service member is less talkative than usual, shows limited response to jokes or cries. He is unable to enjoy relaxation and companionship, even when the tactical situation permits.

Change in Outward Appearance

If the Service member is in a depressed mood, he may be observed to exhibit very little body movement and an almost expressionless (mask-like) face.

Substance Abuse

Some Service members may attempt to use substances such as alcohol or other drugs as a means of escaping combat stress. The use of substances in a combat area makes some Service members less capable of functioning on the job. They are less able to adapt to the tremendous demands placed on them in combat.

Loss of Adaptability

Less common reactions include uncontrolled emotional outbursts such as crying, yelling, or laughing. Some Service members may become withdrawn, silent, and try to isolate themselves. Uncontrolled reactions can appear singly or in combination with a number of other symptoms. In this state, the individual may become restless, unable to keep still, and move aimlessly about. He may feel rage or fear, which he demonstrates by aggressive acts, angry outbursts or irritability.

Disruptive Reactions

Service members with disruptive, combat stress reactions—

- **Cannot function on the job.** In some cases, stress produces symptoms often associated with head injuries. For example, the person may appear dazed and may wander around aimlessly. He may appear confused and disoriented, and exhibit either a complete or partial memory loss. Service members exhibiting this behavior should be removed from duties until the cause for this behavior can be determined.

- **Compromise their own safety.** In a desperate attempt to escape the danger that has overwhelmed him, an individual may panic and become confused. The term *panic run* refers to a person rushing about without self-control. In combat, such a Service member easily compromises his safety and could possibly get killed. His mental ability becomes impaired to the degree that he cannot think clearly or follow simple commands. He stands up in a firefight because his judgment is clouded and he cannot understand the likely consequences of his behavior. He loses his ability to move and seems paralyzed.
- **Compromise the safety of others.** If panic is not quelled early, it can easily spread to others. A person in panic is virtually out of control and needs to be protected from himself. More than one person may be needed to exert control over the individual experiencing panic. However, it is also important to avoid threatening actions, such as striking him.

Referral to the Chaplain, Medical Officer or Combat Stress Control Team

Although the more serious or warning behaviors described in the preceding paragraphs usually diminish with help from comrades and small unit leaders, and time, some do not. An individual usually improves when able to get warm food, rest and an opportunity to share his feelings with comrades or small unit leader. If the symptoms endanger the individual, others or the mission, or if they do not improve within a day or two, or seem to worsen, get the individual to talk with the unit chaplain or medical officer. Access to mental health/combat stress control specialists may be sought, if available. **Do not** wait too long to see if the experience is better with time. Specialized training is not required to recognize severe stress reactions. The small-unit leader can usually determine if the individual is not performing his duties normally, not taking care of himself, behaving in an unusual fashion, or acting out of character.

1005. SUICIDE AWARENESS

Some behaviors and symptoms previously described are not only signs of stress reaction, but can also signal potential suicide risks. Service members must be ever vigilant for the signs and signals of a potential threat of suicide given by their fellow Service members. Appendix A provides useful tools for identifying these signals.

Chapter 2

Combat Stress Prevention, Management, and Control

2001. PREVENTING AND MANAGING COMBAT STRESS

The same leadership skills that apply to troop welfare and warfighting can effectively reduce or prevent combat stress reactions. Leaders should take preventive actions and address stress symptoms as they appear. Ignoring the early warning signs can increase the severity of stress reactions. Positive action to reduce combat stress also helps Service members cope with normal, everyday situations and makes them less likely to experience harmful combat stress reactions. Table 2-1 lists stress management techniques.

2002. STRESS-REDUCTION TECHNIQUES FOR LEADERS

To reduce stress, the leader should—

- Lead by inspiration, not fear or intimidation.
- Initiate and support stress management programs.
- Provide information to focus stress positively.
- Ensure each Service member has mastered at least two stress coping (relaxation) techniques, a slow one for deep relaxation and a quick one for on the job.

Table 2-1. Stress Management Techniques.

Assure every effort is made to provide for the troops' welfare.
Instill confidence in each Service member and his equipment, unit, and leadership.
Be decisive and assertive; demonstrate competence and fair leadership.
Provide sleep and/or rest, especially during continuous operations, whenever possible.
Ensure sleep for decisionmaking personnel.
Set realistic goals for progressive development of the individual and team.
Systematically test the achievement of these goals.
Recognize that battle duration and intensity increase stress.
Be aware of environmental stressors such as light level, temperature, and precipitation.
Recognize that individuals and units react differently to the same stress.
Learn the signs of stress in yourself and others.
Recognize that fear is a normal part of combat stress.
Rest minor stress casualties briefly, keeping them with their unit.
Be aware of background stress sources prior to combat; e.g., family concerns and/or separation, economic problems.
Provide an upward, downward, and lateral information flow to minimize stress due to a lack of communication.

Table 2-1. Stress Management Techniques (Continued).

Practice stress control through cross-training, task allocation, tasks matching, and task sharing.
Look for stress signs and a decreased ability to tolerate stress.
Practice and master stress-coping techniques.
Face combat stress; it is unhealthy to deny the stresses of combat.

- Look out for Service members' welfare.
- Communicate with Service members personally to learn of stressors and detect signs of stress.
- Understand that stress in response to threatening or uncertain situations is normal.
- Create a spirit to win under stress.
- Act as role model for self-control of stress reactions.

Realistic training is the primary stress-reduction technique. It assures Service members' maximum confidence in their skills and belief that their leaders are doing their best for them. Since the basic necessities of life assume even greater importance on the battlefield, leaders should:

- Ensure personnel are properly trained.
- Ensure training includes understanding of combat stress and how to deal with it.
- Place welfare of subordinates before personal welfare, but keep them capable.
- Ensure Service members get as much rest as possible.

- Ensure the best possible shelters are available.
- Keep Service members well supplied with food, water, and other essentials.
- Provide mail, news, and information avenues.
- Provide the best medical, logistical, and other support.
- Maintain high morale, unit identity, and esprit de corps.
- Keep the same unit members together.
- Assure that experienced unit members take care of and teach new members.

The unit's encouragement and support of Service members' efforts to cope with stress have a decisive effect. Unit actions can determine if Service members will endure combat exposure and accomplish their mission. Stress-coping indoctrination is part of every unit's combat training and is followed by a program of action. Controlling stress requires practice. Programs are tailored to individual units and improve with experience. Coping with stress is practiced under conditions as similar to combat as possible. A unit's ability to cope with stress in combat depends primarily on how rigorously and realistically training has been conducted.

2003. PERFORMANCE DEGRADATION PREVENTIVE MEASURES

Every Service member, team, and unit must learn to effectively sustain performance in continuous operations. This requirement applies especially to leaders. While it is an important ingredient, the determination to endure does not ensure effectiveness. Gaining the required capability goes beyond a high level of proficiency in combat skills and technical specialties. It means learning to identify the adverse conditions of continuous operations, cope with them, and overcome their effects. It also means learning how to slow the rate of performance degradation. Units, leaders, and personnel must prepare and execute plans and train to sustain performance. Adverse conditions progressively degrade Service member effectiveness. Fortunately, long-term remedies exist for slowing the rate of performance decline. These remedies, which must be introduced prior to combat, include safety, food intake, combat load, and physical fitness.

Safety

Safety, which encompasses such factors as using proper lifting techniques and staying alert and careful, is influenced by fatigue. Overtired Service members are more vulnerable to injury than those who are rested. After 72 hours of continuous combat, the tendency to seek shortcuts is very strong, and accident rates increase 50 percent. Fatigue affects all military systems, but it is especially hazardous when weapon systems are involved. Catastrophic accidents can occur when fatigued (and under-experienced) crews man weapon systems. Ways to safeguard Service members include developing and following safety standing operating procedures and increasing supervision during extended operations.

Food Intake

If Service members are too busy, stressed or tired to eat adequate rations during continuous operations, their caloric intake will be reduced. This may lead to both physical and mental fatigue and degraded performance. For example, in accidents judged to involve aviator fatigue, there is some indication that before the accidents occurred, the pilots had irregular eating schedules or missed one or more meals. In field tests done by the U.S. Army Natick Research, Development, and Evaluation Center, Natick,

MA, some personnel subsisting solely on meals, ready to eat (MREs) lost weight over just a few weeks. Leaders must encourage troops to eat all of the main items in their MREs, not just the candy, to get balanced nutrition. In various other field tests conducted during continuous operations, it was discovered that meals were frequently delivered late or missed altogether. Although the relationship between performance and nutrition is not clear, eating regularly is important in continuous operations, as well as providing hot meals at assigned times or when the workload has been sustained boosts morale.

Good nutrition is important. An inadequate diet degrades performance, reduces resistance to disease, and prolongs recuperation from illness and injury. When food and water become available in continuous operations, leaders must provide and maintain a supply of food that has the nutritional value commensurate with the physical activity and stress of battle. They must remind and encourage Service members to eat and drink properly. The excitement, stress, and rapid pace of events associated with field preparations can cause Service members to forget to drink liquids. Thus, they enter the early part of the field scenario inadequately hydrated. Dehydration may result, especially if the early scenario calls for assault of a position or rapid air/land deployment. Contributing to developing dehydration is the relative lack of moisture in MREs and other packets. In addition, Service members experiencing dehydration lose their appetite and reduce their food intake. This, in combination with dehydration, leads to degraded performance. Leaders must reemphasize drinking regimens to ensure that Service members are properly hydrated going into battle. Leaders must remind Service members to drink liquids in both hot and cold climates and must monitor fluid intake. If personnel drink only when thirsty, they will become dehydrated.

Combat Load

In combat, the load carried by a Service member significantly exceeds optimum recommended weights. In the case of a light infantry Service member, the combat load is as much as double the recommended load. Physical conditioning cannot compensate for this degree of excess. Service members tire faster and, in continuous combat, recovery from fatigue becomes more time-consuming.

When the Service member must carry excessive amounts of equipment, the effects of stress and lack of rest are magnified. Employing a load echelonment concept must be considered to ease the strain on Service members. In this concept, the unit separates an individual's equipment into two loads—fighting and existence. As the unit closes the objective, the heavier existence load is dropped and the Service member continues with the lighter fighting load.

Physical Fitness

Good physical conditioning delays fatigue, builds confidence, and shortens recovery times from illness and injury. It also prepares individuals to better cope with the physiological demands of stress. Service members in top physical condition can better control their internal physiological functions, which will improve their overall performance. Physical fitness—including aerobic fitness, muscular strength, and endurance—must be developed in all Service members to strengthen their ability to rebound from exhaustion. Aerobic fitness increases work capacity and the ability to withstand stress. While feelings of depression and moodiness accompany tiredness, aerobically fit Service members are affected less than those unfit. The ability to quickly recover from physically strenuous workloads is maintained by smart physical training, performed consistently and routinely. However, there is no evidence that good physical conditioning significantly reduces normal sleep requirements nor compensates for the deleterious impact of sleep deprivation on cognitive functioning. Sleep deprivation is discussed in greater detail in chapter 4.

2004. EFFECTIVE LEADERSHIP

The effective leader in combat is competent and reliable. He knows his job without question, and he can be counted on to do it regardless of the situation or circumstances.

Effective small-unit leadership reduces the impact of stress in several ways. Leaders understand the sources of combat stress and reactions to them. In addition, leaders manage stress problems to keep them from spreading throughout the organization by implementing the following actions:

- Continue mission performance; focus on immediate mission.
- Expect Service members to perform assigned duties.
- Remain calm, directive, and in control at all times.
- Let Service members know their reactions are normal and there is nothing seriously wrong with them.
- Keep Service members productive (when not resting) through recreational activities, equipment maintenance, and training to preserve perishable skills.
- Ensure Service members maintain good personal hygiene.
- Ensure Service members eat, drink, and sleep as soon as possible.
- Let the Service members express their thoughts. Do not ignore or make light of expressions of grief or worry. Give practical advice and put emotions into perspective.

A battalion commander in World War II made the following address to his Marines just before a new offensive:

> I know as well as you do that the going has been tough. Maybe I sweat it more than you do because I have more to worry about. But you've done a fine job, and I'm proud of every single man in my outfit. I assure you that everything possible will be done to give you the best available support, and I will not order you to attack unless I'm confident that you have a real chance to succeed . . . . The harder we fight now, the sooner we can finish this mess and get back to living the way we want.

That speech demonstrated a common sense approach in dealing with stress. Had this commander preached about the advantages of democracy over fascism or the evils of Hitler, his words would have fallen on deaf ears. Instead, he seized on the strong motivators of his troops: their *respect* for him as a leader, their desire to continue to be worthy of his respect, and their desire to share in the unit's accomplishments.

Generally, confidence is based on the technical and tactical competence exhibited by leaders and Service members. Since confidence is one of the strongest defenses against stress, its development is a major goal for every military leader. To gain confidence, Service members must believe in themselves, their equipment, other unit members, and their training. Above all, they must believe in their leaders' competence. Each of these beliefs is instilled and reinforced at every opportunity.

The following comments were made by a Marine colonel reflecting on his experience as a company commander in Vietnam.

> A feeling of helplessness will overtake men when they realize that they have been put in an untenable position. When the situation is so bad that the men cannot fight back, they are susceptible to fear. The

situation can be critical, but as long as the men can fight back, fear normally will not overtake them. The solution to this problem is: don't put your men in untenable positions. Granted, in every battle some men in a unit will find themselves in a situation where they can't fight back. This is inevitable especially in an attack. Proper use of fire team, squad, platoon, and company formations, as well as intelligent use of terrain, minimizes the danger of a large portion of the unit being pinned down to the extent where the men cannot return fire. If you are tactically and technically proficient, use common sense, apply the principles of war, and employ the firepower available to your unit, untenable situations can be avoided. Simply stated, know your job!

Those comments provide an insight and sensitivity to problems related to combat stress through small-unit leadership. His instincts told him how important it was to exhibit solid leadership qualities in order to allow his troops to operate at maximum effiiency. A unit builds confidence, esprit, integrity and cohesion when the *leaders know their jobs*.

2005. COMBAT STRESS MANAGEMENT BUILDING BLOCKS

Confidence in Leaders

Leaders must demonstrate effective leadership to earn their subordinates' loyalty and trust. Leaders are responsible for—

- Committing the unit to missions commensurate with abilities.
- Planning operations carefully and thoroughly.
- Preparing the unit to accomplish the mission.
- Leading and guiding the unit to mission accomplishment.

- Showing consistent good leadership that convinces subordinates their leaders know best what should be done, how it should be done, who should do it, and how long the task should take. Authority accompanies leadership beyond the automatic authority given by military rank and position. Authority and respect are earned based on confidence in a leader's ability to guide the unit to success.

Confidence in Training

Training helps Service members develop the skills required to do their jobs. Confidence is the result of knowing they have received the best possible training for combat, and are fully prepared. This confidence results from the following:

- Realistic training that ends with successful mastery.
- Relevance of training to survival and success on the integrated battlefield.
- Refresher and cross training.
- Systematic individual and collective training.

Confidence in Unit

Each Service member in a unit needs to become confident of the other unit members' competence. Individuals must stay and train together to gain that personal trust. Unless absolutely necessary, teams should not be disbanded or scrambled. Subunits in the same larger unit should have the same Standing Operating Procedures and training standards, so members can fit in quickly if teams have to be cross-leveled or reorganized after casualties occur. Confidence in the unit leads to feelings of security, which in turn allows members to sleep and positively focus stress. In combat, the unit must receive each member's highest commitment to unit loyalty. Mission accomplishment is the unit's highest priority.

Confidence in Equipment

Service members who learn to operate and maintain assigned equipment develop confidence in their ability to employ it. This, in combination with an individual's belief in his personal capabilities, raises overall confidence in fighting ability.

Cohesion and Morale

Good unit cohesion and morale offset the negative effects of combat stress. The foundation for any stress reduction program includes trust and confidence in the following:

- Fellow Service members.
- Competence and fairness of the unit leaders.
- Unit's technical abilities and military power.
- Equipment.
- Personal combat ability.
- Sense of support from the civilian community.
- Personal spiritual well-being.

Physical Conditioning

A strong relationship exists between physical stamina and the ability to resist combat stress. Good physical conditioning has physical and psychological benefits. Rigorous physical conditioning helps protect against the stress of continuous operations. A regular program of physical fitness to increase aerobic endurance, muscular strength, and flexibility is essential to combat readiness. As physical conditioning improves, Service members feel better about themselves, have greater confidence in each other, and their stress is reduced.

Unit training includes regular physical conditioning. This increases the members' tolerance to all types of stressors. The program is geared to the unit's combat mission, and exercises are tailored to the environment where the unit operates. The pace, length, and types of runs, road marches, and other activities are commensurate with the unit's need. Light infantry units need more demanding, longer road marches than maintenance units. Activities also include team athletics, which capitalize on cohesion-building aspects as well as physical benefits. The benefits of such a program include developing endurance through aerobic exercises, enhancing strength through weight training, and deprivation/physical stress training.

Family Care

Service members entering combat with financial worries or family problems risk breaking down under the additional stress of combat. Even positive but unfinished changes on the home front, such as a recent marriage or parenthood, can distract the Service members' focus on combat missions with worries that they will not live to fulfill their new responsibilities at home.

Leaders must be aware of this risk and assist members in handling personal matters before deployment. Pre-deployment and post-deployment family briefings should be conducted, and programs established to assist families before, during, and after deployment. When Service members know their families are cared for, they are better able to focus on their combat duties.

Coping with Individual Stress

Stress pushes the body to its limits and causes tension; relaxation reverses this process. Coping with personal stress is essential. Stress-coping skills are incorporated into unit training early, and command emphasis is placed on practicing them. Service members receive a block of instruction on stress-coping techniques, then are given supervised time each duty day to practice them quietly. After 3 or 4 weeks of practice, most Service members relax easily and quickly, even under highly stressful conditions. They will be able to naturally control stomach fluttering, heart rate, blood pressure, and stress.

The stress-coping exercises include deep breathing, muscle relaxation, and cognitive exercises. Deep breathing is the simplest to learn and practice; the others require longer instruction and more practice time. On request, the Combat Stress Control (CSC) team will provide instructional materials and assistance.

Deep-Breathing Exercise

This consists of slow, deep inhaling, which expands the chest and abdomen, holding it for 2 to 5 seconds, then exhaling slowly and completely, which pushes out the used air. This can be done for five breaths as a quick, mind-clearing exercise, or continuously to promote sleep. Abdominal or diaphragmatic breathing (making the stomach move the air, rather than the upper chest) is especially effective for stress control and, with practice, can be done simultaneously with task that require full attention.

Muscle Relaxation Exercises

These are more complex. They generally consist of concentrating on various muscle groups and the tensing and relaxing of limbs to relax the entire body. Quick versions for use in action consist of tensing all muscles simultaneously, holding for 15 seconds or more, then letting them relax and "shaking out the tension." Deep relaxation versions start in the feet and work up (or start in the head and work down), body part by body part (muscle group by muscle group), tensing and then relaxing each in turn, while noticing how each part feels warm after it relaxes.

Cognitive Exercises

These consist of self suggestion (positive self-talk); imagery (imagine being fully immersed in a deeply relaxing setting); rehearsal (imagine performing the stressful or critical task under pressure and doing it perfectly); and meditation (clearing the mind of all other thoughts by focusing on every breath and silently repeating a single word or phrase). These techniques involve creating positive mental images that reduce the effects of stressful surroundings, redirecting mental focus and learning to detach from stress. Service members are encouraged to practice stress management techniques and discuss their use in combat.

CAUTION

It is important not to use deep relaxation techniques at times when you need to be alert to dangers in your surroundings. Practice the quick relaxation techniques so you can use them automatically without distraction from the mission.

2006. CRITICAL EVENT DEBRIEF AND UNIT LEADER'S AFTER ACTION REVIEW

NOTE: On the subjects of Critical Event Debrief/After-Action Review, U.S. Army personnel will follow the guidance provided in FM 22-51, *Leaders' Manual for Combat Stress Control*, chapter 6.

Critical Event Debrief

A Critical Event Debrief (CED) is a structured group process designed to mitigate the impact of a critical event and to accelerate normal recovery of those personnel involved. The CED is normally conducted by a team composed of trained members, (medical officers, chaplains, mental health professionals, trained unit members). A CED's main value is to quickly restore unit cohesion and readiness to return to action, through clarifying what actually happened and clearing up harmful misperceptions and misunderstandings. It may also reduce the possibility of long term distress through sharing and acceptance of thoughts, feelings, and reactions related to the critical event. Ideally, CEDs are conducted 24 to 72 hours after the occurrence of the event, away from the scene and separate from any operational debriefing. The typical CED lasts from 2 to 3 hours. Situations that warrant a CED include:

- Death of unit member
- Death or suffering of noncombatants (especially women and/or children)
- Handling of the dead, management of carnage or the sight of devastation during disaster work
- Serious friendly fire incident

- Situation involving a serious error, injustice or atrocity
- Terrorist attack.

After Action Review

Although a CED would be warranted following such traumatic events as those listed, if access to CED trained professional team is not available, small-unit leaders can modify their After Action Review (AAR) to assist in identifying the level of stress reaction of unit members. The AAR or "hotwash" is a routine practice used by leaders to debrief post-mission operations. Accordingly, it provides a familiar, non-threatening forum for leaders to identify the levels of stress reaction experienced by the unit members using the information in paragraph 1004. The leader's AAR may be sufficient to restore unit readiness for further action. It may also identify personnel who need immediate or later referral for chaplain or mental health/CSC help for combat stress-related symptoms. In those instances when access to CED trained facilitators is not available, the unit leader should incorporate the following questions in the AAR:

Fact Phase
Ask unit members to describe the event from their individual perspectives. What was each member's specific role in the event?

Thought Phase
What were each member's first thoughts at the scene (or when the incident was first heard of)?

Reaction Phase
What was the worst thing about the event? What was thought and/or felt?

Symptom Phase
Describe probable thinking, physical, and emotional responses both at the scene and a few days afterward.

Training Phase
Relay information regarding stress reactions and what can be done about them. If prepared handouts are available, distribute them. Include points of contact within the unit.

Wrap-up Phrase
Reaffirm positive things. Summarize. Be available and accessible. Debriefing team/unit leaders together decide which individuals need more help or referral.

2007. EFFECTIVE COMBAT STRESS CONTROL PROGRAM

An effective stress management program starts with early planning, continues during employment, and extends beyond the return home. The CSC program is comprised of three phases: pre-deployment, deployment and combat, and post combat.

Phase 1: Predeployment

During pre-deployment, some stressors facing Service members include long working hours, preparation for training, fear of the future, family worries, and anxiety about the unit's readiness. Signs of poor coping include insomnia, increased use of alcohol, marital problems, increased bickering in the unit, irritability, and suicidal feelings. The most important preparatory steps to take in the predeployment phase are to:

- Conduct unit training and mission rehearsals.
- Prepare for changed sleep schedules and jetlag.

- Attend to task assignments and allocations.
- Conduct equipment and supply maintenance checks.
- Attend to personal and family matters. (Call the Family Service Center.)
- Integrate new members into the unit positively and actively.
- Welcome "significant others" (not just entitled beneficiaries) in the Family Support network information tree.
- Brief as much information about the operation as possible, consistent with operations security measures.
- Actively familiarize the unit members with the stressors and horrors they may encounter (for example, watch videotapes from network television showing suffering civilians or atrocities), but always followed immediately with what the unit will do in those situations.
- Arrange for pre-deployment training and education, especially for refresher training of stress reduction techniques from local mental health professionals, a Special Psychiatric Rapid Intervention Team (SPRINT) or CSC team, if available.

Unit Training

Because unit leaders have combatted the stressors associated with garrison living and peacetime training, they have learned to know their Service members and what affects their performance. All Service members should believe unit training has prepared them thoroughly for combat. Stress is reduced if the training improves belief in themselves, and the unit's ability to fight and win. They are assured training is designed and developed to meet combat requirements, provides the ability to fight successfully, and that supporting units receive the same realistic training.

Realistic mission rehearsal helps desensitize Service members against potential combat stressors. For example, wearing and realistically training in protective gear is important. By doing so in pre-deployment training, Service members become less distressed in combat. The unit should be exposed to extensive drills wearing protective gear during live-fire exercises before employment. Such training has the two-fold, stress-reduction benefit of building confidence and preparing Service members for combat.

Leaders who provide their Service members with advance knowledge about the enemy prepare them for the stress created during deployment and initial enemy encounter. It is important during such training to talk realistically about enemy strengths and weaknesses as well as those of their own units. While inspirational pep talks are also important at this time, they should not include biased, inaccurate information. Leaders earn trust and respect if their troops perceive them as accurate, dependable sources of information.

Physical Fitness Training

In the time-urgent work of preparing for deployment, leaders assure that physical fitness is sustained, either by the work involved or by selective unit physical training exercises. New unit members who are insufficiently fit receive special training.

Stress-Coping Skills Training

During preparation for deployment, the leaders direct the unit to practice stress coping and relaxation techniques, and are positive role models by demonstrating use of these techniques. If necessary, the chaplain and mental health professional personnel available to the units can provide additional training.

Sleep Discipline

Before deployment, unit leaders must consider fatigue and sleep loss occurring during combat. The enforcement of work and rest schedules begins early in pre-deployment training. During continuous operations, fatigue caused by lack of sleep is a major source of stress. Breaks in combat are irregular, infrequent, and unscheduled. Extended sleep is unlikely. *Sleep logistics* is emphasized such that sleep and rest are allocated or supplied like rations, water, equipment, and ammunition. Sleep discipline training addresses the following points:

- A unit-specific work-rest-sleep plan is developed and practiced.
- The unit leader or commander is included in the allocation of sleep and rest time, as lack of sleep will impair his judgment and decision-making skills as much as those of his subordinates.
- The plan allows Service members at least 5 hours of uninterrupted sleep, ideally between 2400 and 0600, every 24 hours. Persons receiving only 5 hours per 24 hours over a period of several days will accumulate a significant sleep debt.
- Sleep priority is assigned to those whose judgment and decision-making are critical to mission accomplishment.
- If received frequently, 1 hour of sleep or even 15-minute naps help, but "slow mental starting" upon wake up can result.
- Relaxation exercises complement sleep schedules. These exercises are used as an alternative to regular sleep or as an aid to help Service members rest under difficult circumstances.

If preparation is made for jetlag, its effects are minimal. The ideal method is to seclude units up to several weeks before deployment, gradually changing work, eat, and rest cycles to coincide with the destination's schedule. Although rapid deployment does not permit this method, there are ways to help make the transition to a new environment, such as changing to new sleep cycles collectively versus individually; sleeping up to 12 hours before operation start; ensuring leaders receive rest; and adjusting diet to meet energy needs. If the unit is divided into shifts, make sure that shift assignments are maintained from day to day, i.e., if a member is assigned to the 0600-1400 shift, keep him on that same shift each day.

Task Allocation and Management

Overloading Service members with tasks or responsibilities is another major source of stress. Allocating tasks fairly among available Service members improves unit effectiveness as well as decreases stress. Proper allocation of tasks include:

Selecting the Right Person for the Job

The right person is fitted to the right task according to the task requirement and the individual's talents, abilities, and training.

Duplicating Critical Tasks

Two Service members are assigned to a critical task requiring mental alertness and complete accuracy. They check each other's work by performing the same task independently.

Cross-Training

Each Service member is trained in a secondary duty position to ensure competently stepping into the position of another.

Developing Performance Supports

Develop standing operating procedures, checklists or other mental aids to simplify critical tasks during periods of low alertness.

Equipment Maintenance and Supply

During pre-deployment, the unit maintains its equipment and manages needed supplies. Once deployed for combat, Service members require confidence that supplies are ample and equipment is dependable. The following questions are important:

- Does the unit provide ample training in equipment maintenance and troubleshooting?
- Has the unit's equipment been field-tested under realistic conditions? For example, have Service members fired and cleaned their weapons while wearing full combat gear or protective clothing?
- Does the unit have sufficient ammunition, food, water, and other essential supplies?
- Does the unit have contingency plans for procuring and managing critical supplies if normal channels are disrupted?

Personal and Family Matters

Family stress adds to combat-imposed stress and causes distraction, interference with performance of essential duties, and a negative impact on stress-coping ability. This will result in the unit's inability to perform at peak. The unit should help the Service members resolve important family care matters before deployment and develop methods for helping families when Service members are deployed. Service members are encouraged to:

- Generate or update their wills.
- Finalize power of attorney for spouses.
- Update life insurance policies, including Servicemen's Group Life Insurance (SGLI).
- Ensure family automobiles are in good repair.

- Develop lists of telephone numbers of reliable points of contact for specific problems (mechanics, emergency transportation, babysitters, sources of emergency money, health care, etc.).
- Resolve major legal issues such as alimony payments, property settlements following divorces, and child support payments.

Role of Leaders

Unit commanders should:

- Brief families as a group before deployment or as soon as possible after deployment. Within the bounds of security limits, explain the mission's nature. Even if a mission is highly confidential, families benefit from such a meeting by being told of the support available to them while separated. They begin to solve problems and form support systems with other families. This includes an opportunity to discuss family questions and concerns. The Family Services Office, base mental health service or the chaplain's office assist in staging this briefing.
- Establish points of contact (e.g. Key Volunteer Network) to assist with family problems. These volunteers possess good working relations with the chaplain and mental health personnel to assist with the management of complex problems.
- Establish key volunteer communication and support networks. Commanding officers' spouses, or spouses of sergeants major are often good resources for developing and running such networks; however, the involvement of junior Service members' spouses is also crucial. Some of the most enthusiastic participants are tasked to make *outreach* visits and encourage shy or depressed spouses to participate.
- Have mental health professionals conduct meetings to discuss post-deployment problems. For example, some children have

difficulty adjusting to a parent's absence, and other family members may experience similar difficulties. Mental health professionals give families valuable information on these normal reactions and suggest ways to prepare for them.

Phase 2: Deployment and Combat

In addition to the normal stress associated with moving to a combat zone, Service members in this phase start worrying about their survival and performance under fire. Their thoughts become centered on fear of the unknown. Leaders should emphasize that stress under these circumstances and conditions is expected and is a natural reaction. This will help prevent "normal" stress reactions from escalating into extreme reactions.

Unit leaders should provide as much information as necessary to their survival and mission success, reinforce the stress control techniques learned during pre-deployment, and help their subordinates understand what happens to them when stressors occur.

Deployment Vehicle

The deployment vehicle—in most cases, an airplane—is a stressor by itself. If it is a commercial aircraft, in-flight problems are usually minor. However, if the unit deploys on a military aircraft, leaders should accomplish the following:

- Designate areas for light exercise and stretching to counter seating discomfort caused by a buildup of leg fluids.
- Ensure Service members drink enough fluids to prevent dehydration and have access to head/latrine.
- Adopt the activity schedule of the new time zone. If the unit is in the sleep cycle or is already in or about to enter the sleep

cycle, cover windows; reduce lighting; and issue earplugs, blankets, and pillows.

- Allow uninterrupted sleep. If a stopover occurs during a sleep cycle, do not waken Service members to eat or partake in activities. If the stopover occurs during an activity period, take full advantage of it by having Service members take washcloth baths, stretch, and perform head-and-shoulder rotations.
- Upon arrival in the area of operations, follow the schedule of the new time zone. Eat the next meal and go to bed on the new schedule. Doing so helps the Service members' bodies adjust.

Information Flow

Since uncertainty about the future is a major source of stress, timely and accurate information becomes vital. Lines of communications are clearly defined and kept open. Issuing warning, operation, and fragmentary orders is critical to ensuring adequate information flow. Informational meetings are conducted at regular intervals, even when there is no new information to disseminate. This reinforces the organizational structure and the importance of unit meetings as the source of current, accurate information. Reliable sources of information are especially important for countering rumors.

Service members also need information or *performance feedback* after mission completion. Merely engaging in a firefight or completing a mission is insufficient. Service members must be told how they performed as a group. The knowledge of mission accomplishment and progress builds unit cohesion, develops a winning attitude, and reduces the effects of stress.

Family Support

Marine Corps Family Team Building programs and corresponding programs for other Services provide family support throughout deployments. The Navy-Marine Corps Relief Society, Army Emergency Relief, American Red Cross, Family Service Centers, and other community agencies also provide direct assistance to family members. The Key Volunteer Network and the American Red Cross continue to function as conduits for emergency information between Service members and their families. Leaders need to educate Service members about these programs and agencies that are available to serve the needs of the community. Effective communication and caring support networks help to prevent anxiety while Service members are deployed and/or in combat.

Religious Ministry Support and Pastoral Care

The Religious Ministry Team is an integrated part of the battalion or unit. In the Marine Corps, a Religious Ministry Team (RMT) consists of a chaplain and Religious Program Specialist (RP) or chaplain's assistant. RMTs are organic to commands or are assigned by higher headquarters to provide direct religious ministry support to a command. (**NOTE:** In the Army, the same teams are called Unit Ministry Teams (UMT).) During operations, the RMT often travels with the forward battalion aid station (BAS) in order to minister to the wounded and to be with the Service members who are most likely to experience combat stress. Using their professional training, skills, knowledge, and relationship with the Service members, chaplains provide care focusing on prevention of mild and moderate combat stress reactions. The RMT's primary mission is to provide ministry and pastoral care to the troops, offering faith, assurance, and hope. Before, during, and after the mission, the team provides field services, sacraments, and counsel, which bring comfort, assurance, and encouragement to Service members as they integrate their experiences into their lives.

In addition to being a spiritual/pastoral mentor for Service members, most chaplains are trained in some form of CED process. Chaplains are effective CED team members as well as trainers of small-unit leaders (e.g., platoon leaders, noncommissioned officers (NCOs), corpsmen, etc.) in CED team member skills and stress management techniques. Appendix B explains the roles of the Religious Ministry Teams.

Physical and Recreational Activities

Service members need an outlet for the anger, frustration, hostility, and grief developed in combat. It is unwise to conduct sports and recreation activities under observation by the enemy, but these activities can be conducted further to the rear. Although makeshift athletic games help relieve stress, they should be conducted with caution. It is interesting to note that the most frequent medical problem in Operation Desert Shield/Desert Storm was sports injury. Other activities, such as listening to music, reading or practicing relaxation exercises are encouraged at rest and recuperation facilities. Time and access to personal hygiene items and facilities are psychologically valuable in combat.

Humanitarian assistance and community relations' projects provide excellent opportunities for Service members to constructively engage in meaningful work. Public Affairs Officers, Civil Affairs Officers, and chaplains are excellent personal resources for developing working relationships with local nongovernmental organizations (NGOs) and private voluntary organizations (PVOs) in the coordination of such projects.

Integration of New Unit Members

The arrival of replacements requires small-unit leaders to conduct continuous training programs. Leaders should be concerned with the various stressors affecting new Service members, as well as those Service members who have been around for a while, including seasoned combat veterans. Leaders teach newcomers to use stress coping skills. Replacements are quickly integrated into their units and become thoroughly acquainted with all aspects of the new unit. The unit also becomes thoroughly acquainted with the replacements. A brief orientation with a sincere welcome, with NCOs and officer support, eases replacement transition for combat.

It is important Service members adjust to a new unit quickly and effectively. If they do not feel they are part of the group, unit cohesion and morale suffer, decreasing unit effectiveness. New Service members in garrison are assigned sponsors to assist them, introduce them to the unit, and smooth their families' adjustment. In combat, replacements are placed with suitable groups of seasoned veterans. New unit members are much more likely to become battle or stress casualties than are members of a seasoned and cohesive group. The veterans need to give support and advice to the newcomers, by example and direct action, if necessary.

Phase 3: Post-Combat

Just as pre-deployment and combat are stressful, the period after combat is also difficult. Today's rapid transportation enables Service members to travel from the battlefield to their hometowns in 48 to 72 hours. This short time often does not give them reflection with their comrades. Units should therefore set aside time in the last few days before leaving the theater to conduct their own "End of Tour" debriefing in which they start at pre-deployment and talk about whatever stands out in their memories, good or bad, as they recount the operation up to its end. There should also be appropriate memorial ceremonies and rituals that formally bring the operation to a close. Awards, decorations, and other recognition must be allotted fairly by the commanders.

Unit officers, staff NCOs, and NCOs, assisted by the chaplains and mental health/CSC teams, prepare the Service members for problems encountered during family reunion. For example, most Service members expect to resume roles and responsibilities they had prior to separation. However, their spouses often resist giving up their new roles as decision makers and primary home managers. Also, a spouse may feel that his or her sacrifices during the Service member's absence have gone unrecognized. This feeling becomes an additional source of tension. If at all possible, the families should receive the same briefings or written materials. Families need to be reassured of their contribution. Key Volunteer Networks and other Marine Corps Family Team building programs, and corresponding organizations for other Services continue to help manage problems with reunion and adjustment.

Service members are briefed that startle reactions to sudden noise or movement, combat dreams and nightmares and occasional problems with sleeping, and feeling bored, frustrated and out of place are common when first returning from combat to a peacetime, civilian setting. The leaders, chaplains, and the CSC team emphasize the normalcy of such reactions. Service members are also advised on resources available to help deal with such symptoms, if they are persistent and become upsetting.

Chapter 3

Command Leadership Actions

3001. LEADERSHIP ACTIONS AND INTERVENTIONS FOR COMBAT STRESS

It is important for military leaders to know something about the treatment of combat stress reactions. As most cases of initial combat stress symptoms do not require medical treatment, it has been found that military leaders are often quite adept at treating less severe cases of combat stress. In most cases, debriefing the mission, including any traumatic events, coupled with rest, food, and sleep, is often sufficient to alleviate the symptoms. If the operational tempo permits the Service member to remain with his unit, and he responds to simple reassurance (e.g., "You just need rest, you'll be okay tomorrow"), he is not a casualty (by definition), and may not require further referral for specialized care. Military leaders in combat often make such decisions. When a Service member requires medical attention to rule out a possible serious physical cause for his symptoms, or because his inability to function endangers himself, the unit, and the mission, he should be transported to the BAS or equivalent nearest medical support facility. Interventions at the small-unit level include:

- If a Service member's behavior endangers the mission, himself, or others, the leader should take appropriate measures to control him.
- If a Service member is upset, let him talk about what is upsetting him, listen, then try to reassure him.
- If a Service member's reliability becomes questionable:

- Unload the Service member's weapon.
- Remove the weapon if there is a serious concern.
- Physically restrain the Service member only when safety is a concern or during transport.

- Reassure unit members that the signs are probably a normal combat stress reaction, and will quickly improve.
- If the combat stress reaction signs continue:
  - Get the Service member to a safer place.
  - Do not leave the Service member alone. Keep someone he knows with him.
  - Notify the senior NCO or officer.
  - Have the Service member examined by medical personel.
- If the tactical situation permits, give the Service member simple tasks to do when not sleeping, eating or resting.
- Assure the Service member that he will return to full duty as soon as possible.

The following quote was taken from an article about fear. It was written by a Marine who articulated his experiences during his tours in Vietnam. It provides intricate insight into how many Service members think and bond during combat.

> What should be done when men succumb to fear brought on by the realization that they are going to die? Before discussing this, first let us look at the men themselves. There are very few men who can be classified as cowards. Most men have too much self-respect to let their buddies down. It is the rare man indeed who will willingly violate the trust of his peers. The vast majority of men will give their lives rather than violate this trust. Proper training of combat troops, prudent leaders who are technically and tactically sound, and the reluctance of men to violate the trust of their peers are the foundation

of a solid combat unit. In this environment, courage and sacrifice are the rule, not the exception. At any given moment, anyone can be rendered ineffective by fear if one realizes that he is going to die. When this happens to a man, do not overreact. He knows he has let his buddies down, and he knows that his buddies and leaders know he has let them down. This is a very uncomfortable feeling for a combat Marine. The men around him will not make a big deal about his actions because they understand the situation. To make an issue of this situation will destroy the man and usually alienate the rest of your men. The man will most likely bounce back to his normal performance. The leader must keep in mind that today's coward is tomorrow's hero. If a man does not bounce back and continues to succumb to this type of fear, the Marines in his fire team will let you know when they have given up on him and no longer consider him trustworthy. Situations such as this are rare, so there is no need to make an example of the man. The biggest pitfall to be avoided is to judge the man's actions as a failure of your leadership as this will indeed cause you to overreact.

3002. SAFETY CONSIDERATIONS

Confiscation of a Service member's weapon should only be considered when it is clearly apparent that the Service member is unreliable and a safety hazard to himself or others. Service members identify with other Service members who are trusted and needed by their comrades. This is one of the strongest factors that aid in returning battle-fatigued members to effective duty. The small-unit leader relocates battle-fatigued Service members to a safer, quieter place to rest and work for several hours up to 1 or 2 days. If the unit cannot wait for the Service members to recover, it is necessary to evacuate them to the first level supporting medical unit. From there, every effort is made to move the Service members to a nonmedical unit or area (a tent or building of opportunity could suffice) for rest, replenishment, and reassurance. It should be made clear that the Service members are tired and in need of an opportunity to talk, to sleep, eat, and replenish fluids; they are not patients. Some small detachments without first sergeants or platoon/company organizations are attached to other units for support and rely on the effectiveness of small-unit leaders. Each Service member is accounted for and every effort is made to ensure strong lines of communications are in place and maintained between Service members and their original unit. Key to successful treatment is the return of the Service member to his original unit.

Actions to be taken for severely combat-stressed Service members are the same as those for the moderately combat-stressed, with one exception: medical personnel at the BAS level should evaluate severely combat-stressed Service members as soon as possible. Casualties will be treated and released within hours; held for rest and replenishment; or evacuated for further restoration. Service members who recover from combat stress reactions; return to their original units, i.e., same company or platoon; and are welcomed upon their return are less likely to suffer recurrence. Once rested and returned, they usually become healthy again. New joins and replacements in the unit are more susceptible to succumb to combat stress. Accordingly, this risk is reduced when Service members recovering from combat stress return to the same unit where their combat experience is known and welcomed. In rare instances, however, it is in the best interest of the individual to be reassigned to other jobs or units.

3003. COMBAT STRESS CASUALTY INTERVENTION MODELS

Management of combat stress reactions is unlike the treatment of physical trauma. Severely injured Service members are stabilized as rapidly as possible and then transferred to the rear. In all wars since World War I, combat stress casualties treated in the rear rarely return to full duty. In contrast, when the same casualties are treated near the front, approximately 75 percent return to full duty. Of those returned to full duty, only 10 percent experience continued symptoms requiring further treatment. Some studies suggest half of those treated at the rear go on to have chronic psychiatric symptoms, and approximately half return to full duty. Therefore, it is clearly in the Service member's best interest to be treated at the front and returned to duty. This also serves the best interest of the unit because it receives a veteran Service member back to duty, rather than breaking in a replacement.

Guidelines for dealing with Service members unable to function because of combat stress reaction are summarized in the memory aid **BICEPS.** BICEPS stands for brevity, immediacy, centrality (Marines)/command contact (Army), expectancy, proximity, and simplicity.

- **Brevity.** Critical Event Debriefing should take 2 to 3 hours. Initial rest and replenishment at medical CSC facilities should last no more than 3 or 4 days. Those requiring further treatment are moved to the next level of care. Since many require no further treatment, military commanders expect their Service members to return to duty rapidly.
- **Immediacy.** CSC should be done as soon as possible when operations permit. Intervention is provided as soon as symptoms appear.

- **Centrality (Marines)/Contact (Army).** Service members requiring observation or care beyond the unit level are evacuated to facilities in close proximity to, but separate from the medical or surgical patients at the BAS, surgical support company in a central location (Marines) or forward support/division support or area support medical companies (Army) nearest the service members' unit. It is best to send Service members who cannot continue their mission and require more extensive respite to a central facility other than a hospital, unless no other alternative is possible. The Service member must be encouraged to continue to think of himself as a warfighter, rather than a *patient* or a *sick person*. The chain of command remains directly involved in the Service member's recovery and return to duty. The CSC team coordinates with the unit's leaders to learn whether the overstressed individual was a good performer prior to the combat stress reaction, or whether he was always a marginal or problem performer whom the team would rather see replaced than returned. Whenever possible, representatives of the unit, or messages from the unit, tell the casualty that he is needed and wanted back. The CSC team coordinates with the unit leaders, through unit medical personnel or chaplains, any special advice on how to assure quick reintegration when the Service member returns to his unit.

- **Expectancy.** The individual is explicitly told that he is reacting normally to extreme stress and is expected to recover and return to full duty in a few hours or days. A military leader is extremely effective in this area of treatment. Of all the things said to a Service member suffering from combat stress, the words of his small-unit leader have the greatest impact due to the positive bonding process that occurs during combat. Simple statements from the small-unit leader to the Service member that he is reacting normally to combat stress and is expected back soon have positive impact. Small-unit leaders should tell Service members that their comrades need and expect them to return. When they do return, the unit treats them as every other Service member and expects them to perform well. Service members suffering and recovering from combat stress disorder are no more likely to become overloaded again than are those who have not yet been overloaded. In fact, they are less likely to become overloaded than inexperienced replacements.

- **Proximity.** In mobile war requiring rapid and frequent movement, treatment of many combat stress cases takes place at various battalion or regimental headquarters or logistical units, on light duty, rather than in medical units, whenever possible. This is a key factor and another area where the small-unit leader helps in the treatment. CSC and follow-up care for combat stress casualties are held as close as possible to and maintain close association with the member's unit, and are an integral part of the entire healing process. A visit from a member of the individual's unit during restoration is very effective in keeping a bond with the organization. A Service member suffering from combat stress reaction is having a crisis, and there are two basic elements to that crisis working in opposite directions. On the one hand, the Service member is driven by a strong desire to seek safety and to get out of an intolerable environment. On the other hand, the Service member does not want to let his comrades down. He wants to return to his unit. If a Service member starts to lose contact with his unit when he enters treatment, the impulse to get out of the war and return to safety takes over. He feels that he has failed his comrades and they have already rejected him as unworthy. The potential is for the Service member to become more and more emotionally invested in keeping his symptoms so he can stay in a safe environment. Much of this is done outside the Service member's conscious awareness, but the result is the same. The more out of touch the Service member is with his unit, the less likely he will recover. He is more likely to develop a chronic psychiatric illness and get evacuated from the war. This is one of the essential principles of CSC.

- **Simplicity.** Treatment is kept very simple. CSC is not therapy. Psychotherapy is not done. The goal is to rapidly restore the Service member's coping skills so that he functions and returns to duty again. Sleep, food, water, hygiene, encouragement, work details, and confidence-restoring talk are often all that is needed to restore a Service member to full operational readiness. This can be done in units in reserve positions, logistical units or at medical companies. Every effort is made to reinforce Service members' identity. They are required to wear their uniforms and to keep their helmets, equipment, chemical protective gear, and flak jackets with them. When possible, they are allowed to keep their weapons after the weapons have been cleared. They may serve on guard duty or as members of a standby quick reaction force.

If treated near their units, 65 to 85 percent of combat stress casualties treated in medical CSC facilities return to duty within 1 to 3 days. About 15 to 20 percent more return to duty in 1 to 2 weeks. Only 5 to 10 percent are sent home, and these usually have other problems in addition to combat stress reactions. If evacuated, few combat stressed Service members will return to duty. In fact, many are likely to be permanently disabled.

History shows that it is important to treat combat stress casualties as close to the front as practicable. Everyone is susceptible to the effects of combat stress, regardless of past performance, rank or duty. *Most Service members receiving combat stress control in accordance with the principles of BICEPS return to duty quickly.* The chances of later psychiatric problems are also reduced.

Combat stress reactions are inevitable in combat, but high stress casualties are not. History shows that highly trained and small cohesive units with good leadership have less than 1 such casualty for every 10 to 15 WIA, even in very heavy fighting. This is significantly fewer than the usual 1:3 to 1:5 ratio in high-intensity battle, and the 1:1 or 1:2 ratio in extreme fighting. Knowledge of factors that increase combat stress disorder in the tactical and overall situation encourages small-unit leaders, comrades, and the individual Service member to act towards building unit cohesion, sharing the burden, and reducing stress. The same measures, plus positive adherence to discipline and the law of war, also prevent combat stress-related misconduct that defeats the purpose of the mission. Tough, realistic training that builds confidence and caring within the unit in combat overcomes many adversities associated with combat stress in a combat environment. However, if a member fails to adequately respond to actions taken by the unit, then he should be evacuated for appropriate medical evaluation.

Chapter 4

Sleep Deprivation

4001. CHALLENGES OF SLEEP DEPRIVATION

People accumulate a "sleep debt" (cumulative loss of sleep over time) when they perform under limited sleep conditions. The only corrective measure for satisfying this sleep debt is sleep itself. Military operations, by their demanding nature, create situations where obtaining needed sleep will be difficult or impossible for more than short periods.

Continuous operations are military operations with many pulses of action every day and night, continuing for several days to weeks, which require careful planning and resource allocation to give everyone a minimum of 4 hours sleep in 24. (FM 22-51)

Sustained operations are continuous operations or combat with opportunity for less than 4 hours sleep per 24 hours for significant personnel, which may be brief or fragmented. (FM 22-51)

Accordingly, service members may have opportunities for only limited or fragmented sleep over an extended period. As a result of these periods of sleep loss, several combat tasks are likely to show decreased performance. These tasks include the following:

- Orientation with friendly and enemy forces (knowledge of the squad's location and maintaining camouflage, cover, and concealment).

- Coordination and information processing (coordinating firing with other vehicles and dismounted elements, reporting vehicle readiness, and communicating with the headquarters).
- Combat activity (firing from bounding vehicle, checking the condition of weapons, observing the terrain for enemy presence).
- Force preservation and regrouping (covering disengaging squads, marking the routes between locations, and conducting reconnaissance).
- Command and control activity (directing location repositioning, directing mounted defense, assigning fire zones and targets).

Continuous operations will potentially be more commonplace on the battlefield. In offensive operations, darkness is the time to retain or gain the initiative; while in defensive operations, obstacles can be employed with greater security during darkness. Forces can disengage undetected and threats to close air support lessen. The physical environment changes at night. As the air cools below ground temperature, inversions reduce visibility and hamper radar and radio signals. Conditions are optimal for using chemical weapons. Visual changes also occur. Without the aid of white light, there is no color perception. There is also a decrease in visual clarity, field of view, and depth perception. Targets take longer to engage. Preparation time increases two-fold to six-fold. Simple actions, such as the departure and return of patrols, become more complex and dangerous. Nighttime planning and coordination require greater attention. Navigation, adjusting fire, and munitions and/or target matching are more difficult. Precision is essential, but accuracy has a price. Service members tend to maintain accuracy at the sacrifice of speed. The adverse conditions associated with or generated by continuous ground combat at night will degrade the fighting performance of Service members, teams, and units. The almost complete mechanization of land combat forces and technological advances that permit effective movement at night, during poor weather conditions, and under conditions of limited visibility have largely overcome the reasons for "traditional" pauses in battle, such as darkness, resupply, and regrouping. New technologies have significantly increased the range, reduced the time, and changed the conditions over which battles are fought. For example, day/night-capable vehicles can operate for extended periods without re-supply, but they are limited by a crew's need to sleep. A Service member is not a machine and is, therefore, the weak link in the chain. The equipment can operate longer than the Service member who operates it, as the Service member must have sleep.

Commanders and leaders must ensure that all Service members obtain enough rest to counteract the effects of rapidly shifting from daytime to nighttime duty hours, or to extended work schedules. Implementing countermeasures that are designed to help Service members adapt to continuous operations conditions can satisfy this requirement. Neither leaders nor their subordinates can perform without rest or sleep. The Service member, the unit, and the leader are all affected by continuous operations. Generally at night, the cognitive and physiological resources of Service members are not at their peak, especially after a rapid shift from daytime to nighttime duty hours. Fatigue, fear, feelings of isolation, and loss of confidence may increase.

Non-stop, unrelieved combat operations (sustained operations) with little or no sleep degrade performance and erode mental abilities more rapidly than physical strength and endurance. Information gained from the Army Unit Resiliency Analysis Model shows that even healthy young Service members who eat and drink properly experience a 25 percent loss in mental performance for each successive 24-hour period without sleep. The mental parameters include decisionmaking, reasoning, memory tasks, and computational tasks. The loss may be greater for Service members who are older, less physically fit, or who do not eat and drink properly.

The effects of sustained operations are sometimes hidden and difficult to detect. Units are obviously impaired when Service members are killed or wounded in action or become noncombatant losses. They are further impaired when their troops are too tired to perform their tasks. Unlike individual performance, unit performance does not deteriorate gradually. Units fail catastrophically, with little warning.

A priority for fighting units is to assure that commanders and leaders are rested and able to think clearly. While this is obvious, it is a most difficult lesson for leaders to learn. During combat, commanders must focus on the human factor. They must assess and strengthen their units as they plan and fight battles. They must accurately decipher which units must lead, which must be replaced, where the effort must be reinforced, and where tenacity or audacity and subsequent success can be exploited. When leaders begin to fail, control and direction become ineffective, and the organization disintegrates. No fighting unit can endure when its primary objectives are no longer coordinated. Leaders must also prepare and precondition Service members to survive. It is particularly important that leaders conscientiously plan and implement effective sleep plans, because activities that are most dependent on reasoning, thinking, problem solving, and decision-making are those that suffer most when sleep and rest are neglected.

Some leaders wrongly believe that their round-the-clock presence during an operation is mandatory; they are unwilling to recognize that they, too, are subject to the effects of sleep deprivation. If the unit has been regularly trained according to the mission command philosophy, two benefits accrue. Not only will a leader be confident that in his absence his subordinates will adhere to his intent, but the trust he shows in his subordinates will continue to maintain unit morale and help ease some of the stress of the situation.

In future operations, the battlefield will become increasingly lethal. The threat of nuclear, biological, and chemical weapons will maximize confusion, uncertainty, and stress, which adversely impact our ability to move, shoot, communicate, and sustain. Sleep loss in this type of environment increases an already stressful situation.

4002. EFFECTS OF SUSTAINED OPERATIONS ON PERFORMANCE

A basic rule for continuous operations is planning ahead to avoid sustained operations, and provide members 5 to 6 hours sleep in 24. However, missions or enemy actions sometimes require exceptional exertion for several days with only unpredictable, fragmented sleep—as required in sustained operations. Sustained combat leads to exhaustion and reduction in effective task performance. Even during the first night of combat, normal sleeping habits and routines are abnormal. The Service member feels the effects of fatigue and the pressure of stress from noise, disrupted sleep time, and threat to life. While essential for endurance, sheer determination cannot offset the mounting effects of adverse conditions. Cognitive degradation involving poor decisionmaking begins during and after the first 24 hours of sleep deprivation.

Individual and unit military effectiveness is dependent upon initiative, motivation, physical strength, endurance, and the ability to think clearly, accurately, and quickly. The longer a Service member goes without sleep, the more his thinking slows and becomes confused. Lapses in attention occur, and speed is sacrificed to maintain accuracy. Continuous work declines more rapidly than intermittent work.

Tasks such as requesting fire, integrating range cards, establishing positions, and coordinating squad tactics become more difficult than well-practiced, routine physical tasks, such as loading magazines and marching. Without sleep, Service members can perform the simpler and/or clearer tasks—lifting, digging, and marching—longer than the more complicated or ambiguous tasks such as a fine hand-eye coordination sequence; i.e., tracking a target through a scope.

Sleep loss affects memory, reasoning, mental assessments, decision-making, problem-solving, subsequent actions, and overall effectiveness. While comprehension is accurate, reading speed slows and recall fails. For example, Service members may understand orders when reading them in documents, yet they are forgotten later when required. Individuals will forget or omit assigned tasks more often than they will make errors in carrying them out.

Leaders can expect declining moods, motivation, initiative, planning ability, and preventive maintenance. High motivation will only increase risk, due to impaired performance. Leaders must recognize erratic or unreliable task performance in subordinates, as well as in themselves. Alertness and performance decline gradually with partial sleep deprivation; that is, when sleep is limited to 4 to 5 hours per night. After 5 to 7 days of partial sleep deprivation, alertness and performance decline to the same low levels as those following 2 days of total sleep deprivation. After 48 to 72 hours without sleep, personnel become militarily ineffective.

Adverse Conditions

Continuous combat forces Service members to perform under adverse conditions that cause degradation in performance. Examples of adverse conditions follow.

Low Light Level

The amount of light available for seeing landmarks, targets, and maps is greatly reduced at twilight and night.

Limited Visibility

Smoke, fog, rain, snow, ice, and glare degrade a Service member's ability to see his environment and objects within it, as opposed to situations free of such conditions.

Disrupted Sleep Routines

People are accustomed to being awake or asleep during certain hours of the day or night. Disruption of the normal sleeping schedule causes degraded performance.

Physical Fatigue

Working the muscles faster than they can be supplied with oxygen and fuel rapidly creates "oxygen debt," eventually making these muscles unable to function until the deficits are made up during brief rests.

Sleep Loss

The muscles can continue to function adequately without sleep, but the brain cannot. Increasing sleep debt leads to subtle, but potentially critical, performance failures.

Sleep Loss Indicators

Indications of degraded performance symptoms become more prevalent as sleep debt accumulates. Performance is affected by the hours of wakefulness, tolerance to sleep loss, and the types of mental or physical work. Both mental and physical changes occur, with symptoms varying among individuals. Leaders must observe Service members for the following indications of sleep loss and degraded performance:

- Physical changes in appearance, including vacant stares, bloodshot eyes, pale skin, and poor personal hygiene. Other physical signs of sleep loss include the body swaying when standing, sudden dropping of the chin when sitting, occasional loss of hand-grip strength, walking into obstacles or ditches, low body temperature, slowed heart rate, and slurred speech.

- Mood changes, decreased willingness to work, and diminished performance go hand-in-hand. Service members may experience decreasing levels of energy, alertness, interest in their surroundings, and cheerfulness with a concurrent increase in irritability, negativity, and sleepiness. Some become depressed and apathetic. Others, for a time, can become energized by sleep loss, talk more, and may be more assertive without necessarily maintaining good judgment. Sleepiness and mood changes are not signs of weakness. After long periods of sleep loss, Service members go from being irritable and negative to dull and weary.

- Service members may feel more effort is needed to perform a physical task in the morning than in the afternoon. Exaggerated feelings of physical exertion may lead to work stoppage, especially between 0400 and 0700. During that time, the tendency to fall asleep is considerably more noticeable than other times.

- Both bickering and irritability increase with sleep loss. When Service members argue, it shows that they are still talking to each other and exchanging orders and messages. When arguments cease, especially after a period of increased bickering, Service members may be in a state of mental exhaustion.

- Comprehension and perception slow considerably. Individuals require extended time to understand oral, written or coded information; to find a location on a map and/or chart coordinates; to interpret changes in enemy fire patterns; and to make sense of things seen or heard, especially patterns. They may have difficulty with spot status or damage reports, and may be unable to assess simple tactical situations.

Loss of Concentration

Sleep deprivation causes the attention span to shorten. There is a loss of concentration on the job as dream-like thoughts cause lapses in attention. Leaders should watch for the following:

- Decreased vigilance. Personnel are less alert and fail to detect the appearance of targets, especially in monotonous environments. They may doze off at the wheel of moving vehicles.

- Distorted attention. Service members may imagine seeing things that are not there, e.g., "moving" bushes when in reality there is no such movement. The sleep-deprived brain can also misperceive bushes, rocks, people, vehicles or anything else and see them as something different, in very precise detail. Often the tired brain "sees" what it wishes were there (food, a bed); at other times, these illusions may be animals or other more bizarre things. But when the mind is alert for an enemy, the brain may generate a very convincing, detailed image of the enemy. Sometimes, but not usually, sounds or other sensations may accompany these illusions. They usually last only seconds, but can persist for minutes if not challenged, and rarely have even been "seen" by equally sleep-deprived comrades when told of them. It is essential for sleep-deprived unit members to check out any questionable things they see with their comrades, and to faithfully follow reporting and challenge procedures.

- Inability to concentrate; easily confused. Service members cannot keep their minds on what they are doing. They cannot follow multiple directions nor perform numerical calculations.

- Failure to complete routine tasks. Sleep loss interferes with completing routine individual tasks, such as drying the feet, changing socks or filling canteens when water is available. Tasks such as performing weapons checks may be skipped.

When a Service member cannot recall what he just saw, read, heard or was told by another individual, he is exhibiting a common sign of sleep loss. His memory loss is limited to recent events. For example, a sleep-deprived Service member may forget recent target data elements or recall them incorrectly and have difficulty learning new information.

4003. ACHIEVING SLEEP IN COMBAT

Sleep deprivation produces stress and, therefore, sleep management is important. Sleep management is a combat multiplier. Planned sleep routines are important for keeping the unit, the individual Service members, and the leader himself functioning as required while reducing sleepiness during continuous combat. Since leaders are responsible for planning sleep routines, they need a basic understanding of the physiological and behavioral aspects of sleep and their impact on performance. The following paragraphs provide this information.

Rhythmic Variations

There are rhythmic variations in individual performance based on a predictable physiological and behavioral cycle that comprises about 24 hours. The 24-hour, day-night/work-rest cycle is called the *circadian rhythm*. Because traveling across a half-dozen time zones disrupts the usual relationship in the day-night/work-rest cycle, for a few days Service members are not sleepiest at their usual sleep period of 2400 to 0600, new-locale time. Allowing sleep about 1200 to 1800, new-locale time, will only delay their adaptation to their new locale. Leaders must instruct troops to go to bed between 2400 and 0600 new-local time to establish a new circadian rhythm.

Another example of circadian rhythm is body temperature. Although one's "normal" temperature is 98.6 degrees, this is really an average or midpoint of a daily swing from 96.8 to 100.8 degrees. For someone accustomed to working days and sleeping nights, body temperature would fluctuate approximately as indicated. There is a well-established link between body temperature and sleepiness and/or performance slumps. Performance parallels body temperature. The higher the body temperature, the better the performance. As body temperature decreases, mood and motivation decline with a concurrent increase in sleepiness and fatigue.

Impact upon performance is most pronounced during the *circadian lull*, which is roughly 0200 to 0600 hours. During this time, performance declines about 10 to 15 percent. In sleep-deprived Service members, this decline may reach 35 to 40 percent. If the day-night/work-rest cycle is disrupted, performance suffers because the Service member is sleepy during the new work period and awake during the new sleep period. The body needs several days to adjust to the new schedule. Critical hours for sleep are between 0200 and 0600 when *anchor sleep* (the most beneficial sleep) is taken. The body is at its lowest temperature during this period. This is the best time for sleeping, but not for napping. To prevent sleep inertia, naps should always be taken at times other than the lowest point in body temperature.

Leaders need to calculate the difference in time zones and make the necessary schedule changes. Leaders will need day-and night-fighting teams. Members acclimated to working days and sleeping nights should be scheduled to work nights and sleep days.

Their performance slump/optimal time to sleep would be 2400 to 0600, new-locale time. Deployment, pre-combat, and combat are not usual circumstances. If certain Service members must have an offset circadian timing from the rest of the unit, a special effort must be made to establish their sleeping time. Obviously, troops must sleep whenever possible. If a planned sleep schedule cannot be followed, however, performance is enhanced if sleep coincides with the low point in body temperature.

Adjusting to new circadian rhythms is a slow process, taking 3 to 6 days to come "in phase" with a new schedule. Leaders should devise a sleep schedule that provides for sleep at the same time of day or night every 24 hours. Sleep schedules that provide for sleep at different times of day or night are less valuable and are detrimental to quality sleep and optimal performance.

Sleep Shifts

Staggered work schedules can be set up for two shifts working 4 hours on/4 hours off, 6 hours on/6 hours off, and 12 hours on/12 hours off. See Table 4-1. Each shift follows the same schedule daily. It is better to maintain regular shift schedules than schedules that continually change.

Sleep/Rest Guidelines

Leaders should use the following sleep and/or rest guidelines in this section to enhance individual and the unit performance in continuous operations.

- Know personal tolerance for sleep loss and those under your command; major individual differences are not easily changed. Individuals who are unable to sleep during pre-deployment and deployment stages should be encouraged to practice relaxation exercises (see paragraph 2005).

Table 4-1. Sleep Shifts.

4 HOURS ON/4 HOURS OFF						
Shift	2400-0400	0400-0800	0800-1200	1200-1600	1600-2000	2000-2400
1	SLEEP	DUTY	SLEEP	DUTY	SLEEP	DUTY
2	DUTY	SLEEP	DUTY	SLEEP	DUTY	SLEEP

6 HOURS ON/6 HOURS OFF				
Shift	2400-0600	0600-1200	1200-1800	1800-2400
1	SLEEP	DUTY	SLEEP	DUTY
2	DUTY	SLEEP	DUTY	SLEEP

12 HOURS ON/ 12 HOURS OFF		
Shift	2400-1200	1200-2400
1	SLEEP	DUTY
2	DUTY	SLEEP

- Ensure that Service members fully use their breaks and other opportunities for rest. Encourage them to waste no time in getting to sleep. Undisturbed, prolonged sleep is the most desirable use of rest opportunities. When there has been sleep loss but little physical exertion (e.g., manning communications, operating a radio), mild physical exercise such as walking around when conditions permit, can help maintain alertness.

- Encourage Service members to sleep, not just rest, by creating the most conducive environment possible for sleep: quiet, without interruptions (or earplugs); dimness or darkness (or with eye cover); not overly warm or cold.

- Do not allow personnel to sleep in unsafe conditions. Enforce strict rules designating sleep areas and requiring perimeter guards. Require day and night guides for all vehicles to prevent Service members from being accidentally run over.
- Ensure that Service members follow sleep schedules or routines. The field commander who does not enforce a sleep schedule or routine leads his troops into an environment that increases the opportunity for hazardous conditions to be encountered while in continuous combat. Taking naps is not a sign of low fighting spirit or weakness; it is a sign of foresight.

Measuring Sleep Loss

Sleep loss can be measured by:

- Keeping a sleep and/or activity log. From pre-deployment to post-deployment, log sleep and nap periods. Service members need 4 to 5 hours per 24-hour period; 6 or 7 hours is optimum. If they receive less, the first chance for a long rest period must be used for sleep.
- Observing performance and asking questions. Look for the indications of sleep loss—such as increase in error occurrence, irritability, difficulty understanding information, and attention lapses—with concurrent decreases in initiative, short-term memory, and attention to personal hygiene. Confirm sleep loss by asking the obvious question: "When did you sleep last and how long did you sleep?"

Sleep Loss Alternatives

Ways to overcome performance degradation include:

- Upon signs of diminished performance, find time for members to nap, change routines or rotate jobs (if cross-trained).

- Have the Service members most affected by sleep loss execute a self-paced task.
- Have Service members execute a task as a team, using the buddy system.
- Do not allow Service members to be awakened for meals while in flight to a new location, especially if the time zone of the destination is several hours different than that of point of departure.
- Insist that Service members empty their bladder before going to bed. Awakening to urinate interrupts sleep, and getting in and out of bed may disturb others and interrupt their sleep.
- Allocate sleep by priority. Leaders, on whose decisions mission success and unit survival depend, must get the highest priority and largest allocation of sleep. Second priority is given to Service members that have guard duty and to those whose jobs require them to perform calculations, make judgments, sustain attention, evaluate information, and perform tasks that require a degree of precision and alertness.

4004. SLEEP/REST PLANNING

Sleep/rest planning applies to the pre-deployment, deployment, pre-combat, combat, and post-combat stages of battle.

Pre-Deployment Stage

Using mission-scenario operation guidelines, determine periods available for sleep and the total number of sleep hours possible. Because continuous operations requirements may change, alternate sleep routines should be planned. Become familiar with the area where the combat unit will sleep; For example, somc may have to sleep in mission-oriented protective posture (MOPP) IV. If sleeping in MOPP IV is anticipated in combat, practice it during the pre-deployment stage. Prior experience reduces stress, so practice anticipated sleep routines before continuous operations.

Deployment Stage

Since sleep will be reduced during deployment, follow pre-planned sleep routines. The prudent commander will choose a 4-hour on/4-hour off, 6-hour on/6-hour off, or 12-hour on/12-hour off shifts from the start. Take into account that Service members on night duty will need to sleep during the daytime. Provide night-shift personnel with separate sleeping quarters to avoid disruption of their sleep period.

Pre-Combat Stage

In general, people are most effective during the afternoon and are least effective just before dawn. Without prior adjustment to the new time zone, which naturally occurs in 3 to 5 days, leaders can expect degraded daytime performance. The reason is that 0200 to 0600 hours home-base time is the low point in performance efficiency and should be considered when planning workloads.

Combat Stage

Every effort should be made to avoid situations where all personnel are physically and mentally exhausted simultaneously. Make the most of any lull during the combat phase by sleeping briefly. Complete recovery from sleep loss may not be possible during intense combat, but limited sleep is helpful. Uninterrupted short sleeps of 15 minutes or longer are beneficial to partially recovering alertness. Sleep during the combat stage may be risky, however, because a Service member may wake up feeling groggy, confused, sluggish, and uncoordinated. It may take his brain from several seconds to 15 minutes to "warm up." Individuals differ in how quickly they take to wake up, but it tends to be worse when the body expected to go into deep sleep, and to get worse with increasing sleep loss. Activities that increase circulation of warm blood to the brain, like moderate exercise or drinking a hot beverage, may shorten the start-up time.

Post-Combat Stage

It is important to make up sleep debt, but experts disagree about the amount of recovery time needed. Some say the hours of sleep needed for recovery after sleep deprivation are less than the amount lost. It is well known and documented that lost sleep is not made up hour-for-hour. Most experts agree that immediately following continuous combat, Service members should be allowed to sleep up to 10 hours. Longer sleep periods are not desirable because they cause "sleep drunkenness" and delay in getting back to a normal schedule. After the first sleep period of up to 10 hours, Service members should return to the regular sleep routine. Sleep inertia lasting longer than 5 to 15 minutes and increased sleepiness may occur for as long as a week following sustained combat. Some experts recommend that 4 of the first 8 hours of recovery sleep should be at the 0200 to 0600 sleep time, and they suggest the following guidelines for complete recovery from the effects of sleep loss:

- 12 hours for sleep and rest after 36 to 48 hours of complete sleep loss with light to moderate work load (fatigue may linger for 3 days).
- 24 hours for sleep and rest after 36 to 48 hours of sleep loss with high workload (12 to 16 hours per day).

- 2 to 3 days time off after 72 hours or more of acute sleep loss.
- As much as 5 days for sleep and rest following 96 hours or more of complete sleep loss.

Most experts agree that 10 hours of sleep is the maximum needed, with the additional 2 hours used for rest. It is doubtful that a Service member could continue past 72 hours of wakefulness. Should this occur, a couple of nights with 10 hours of sleep are more beneficial than an excess of 10 hours during one sleep period. If Service members have not slept for 36 to 48 hours or more, they should avoid sleep of less than 2 hours, especially between 0400 and 0600. A too-short sleep period at the wrong time may cause a long period of sleep inertia. After 96 hours of total wakefulness, 4 hours of sleep may provide substantial recovery for the simpler, less-vulnerable tasks. Recovery continues with additional days of 4 hours of sleep per 24 hours. Complex leadership tasks may require longer recovery sleep, but sleep until fully satisfied is not necessary.

Sleep loss alone does not cause permanent health problems, nor does it cause mentally healthy people to become mentally ill. Reduced sleep (from 8 to 4 hours) does not cause physical harm. Hallucinations may occur, but they disappear after recovery sleep. Clinical laboratory tests show that total sleep loss of over a week does not pose serious health problems. It is doubtful that Service members could stay awake for such an extended period, and it is not suggested that Service members try to endure long periods without rest. However, the effects of sleep loss, such as inattentiveness and poor judgment, may be harmful (such as falling asleep at the wheel of a vehicle).

Sleep cannot be stored in our bodies for emergency use. Sleep of more than 7 to 8 hours before deployment does not "store up" excess sleep, but sleep taken immediately before a deployment can prolong activity. Therefore, it is important to begin continuous operations in a rested state. During daytime or early morning naps, many Service members experience vivid dreams as they fall asleep and often wake up frightened. Leaders should inform their troops that this occurrence is both common and normal during daytime sleep. If a single, unbroken period of 4 to 5 hours is not available for sleep, "power naps" of 15 to 30 minutes, although less recuperative, can be taken. Leaders must capitalize on every opportunity for a "power nap." Merely resting by stretching out does not take the place of sleep. Only sleep can satisfy the need for sleep.

Appendix A

Suicide Awareness Information

SIGNS OF SUICIDE

A person contemplating suicide—

- Believes he or she is in a hopeless situation.
- Appears depressed, sad, tearful; may have changes in patterns of sleep and/or appetite.
- May talk about or actually threaten suicide, or may talk about death and dying in a way that strikes the listener as odd.
- May show changes in behavior, appearance or mood.
- May increase or start drug or alcohol use.
- May injure self or engage in risky behavior.
- Abandons planning for the future.
- May start withdrawing from others, including family and close friends.
- May give away possessions.
- May appear apathetic, unmotivated, indifferent.

PREVENTION KEYS

The small-unit leader may be the first to identify the "signals" of a potential suicide-risk Service member. The keys to prevention are to provide aid to persons at suspected risk and follow the acronym AID LIFE.

A	Ask: "Are you thinking about hurting yourself?"
I	Intervene immediately.
D	Do not keep a secret concerning a person at risk.
L	Locate help (NCO, chaplain, corpsman, doctor, nurse, friend, family, crisis line, hospital emergency room).
I	Inform your chain of command of the situation.
F	Find someone to say with the person. Do not leave the person alone.
E	Expedite! Get help immediately. A suicidal person needs the immediate attention of helpers.

Appendix B

Religious Ministry Team's Role

(Adapted from Army Field Manual (FM) 22-51, *A Leader's Manual for Combat Stress Control.*)

This appendix addresses the general role of the ministry team in the commander's program of combat stress control and in battle fatigue ministry. In the Marine Corps ministry teams are called Religious Ministry Teams (RMTs) and consist of chaplains and Religious Program Specialists (RPs) or chaplain's assistants (In the Army they are called Unit Ministry Teams). The RMT is assigned to a command or designated by higher headquarters to be responsible for the direct religious ministry support to the command. RMTs provide professional ministry support to leaders in fulfilling their combat stress identification and intervention responsibilities. RMTs can also assist in training leaders to recognize combat stress symptoms.

EFFECTS OF STRESS

Negative effects of stress can be lessened when members are prepared physically, emotionally, and spiritually prior to combat/operations. RMTs can assist in preparing Service members to manage combat stress with training before and during deployment. This training helps Service members draw upon their personal religious faith, spiritual strength, and values as well as to share strength and confidence during any operation.

RELIGIOUS MINISTRY TEAM RELATIONSHIP

The RMT's established relationship with the command promotes trust with the troops. As integral members of their commands, RMTs are trained and ready to respond to the needs of Service members experiencing combat stress. A person-oriented resource, the chaplain and RP team minister to combat stress casualties, especially Service members having less severe difficulties who have rapid replenishment potential.

SPIRITUAL VALUES

Service members' inner resources are often based on their personal faith, religious beliefs and spiritual values. In combat, Service members may show more interest in their religious beliefs. When religious and spiritual values are challenged during the chaos of combat, Service members may lose sight of inner resources that sustain them. Service members can then become targets of fear, despair, hopelessness, and eventually, combat stress casualties. They are also at risk for committing misconduct stress behaviors. Chaplains are a source of direction and stability to Service members experiencing these dilemmas and seeking to refocus their personal beliefs and spiritual values.

RELIGIOUS MINISTRY TEAM SUPPORT

Ministry teams can provide preventive, immediate, and replenishing spiritual and emotional support and care to Service members experiencing combat stress.

Preventive

The RMT's religious ministry support assists in preventing combat and misconduct stress behaviors through establishment of a presence within the unit. It is important for the RMT to be present with Service members when the unit trains and when it deploys. The chaplain and RP can be a calming influence on members and can help Service members strengthen or regain values important to them. Chaplains can help prevent combat stress and misconduct by:

- Being present with the Service members and deploying with the unit.
- Providing opportunities for private and group prayer and worship.
- Supplying personal religious articles and materials.
- Reading the scriptures with Service members.
- Providing sacraments as the situation allows.
- Counseling Service members, allowing them to work through stress, fear, anxiety, anger, and frustration.
- Visiting Service members in work and living areas.
- Assisting Service members and families prior to deployment with preparation for geographical separation and an uncertain future through Marine Corps Family Team Building (MCFTB) programs that emphasize family strengths. This helps Service members to know that their families are cared for during deployment.

Immediate

RMTs can assist commanders in the identification of Service members experiencing combat stress. Chaplains often work closely with the medical officer and are trained to recognize the signs of combat stress and provide religious support to Service members experiencing combat stress. Chaplains assess spiritual needs and provide the appropriate religious ministry. This may include such things as:

- Presence with the Service member.
- Conversation and counseling, providing opportunities to share fears, hopes, and other thoughts.
- General prayers; prayers for the individual or fallen comrades.
- Rites, sacraments, and ordinances as appropriate.
- Reading from scriptures.

Replenishing

Following an engagement, the unit may require reconstitution through the addition of new personnel. The RMT can assist surviving Service members rebuild their emotional, psychological, and spiritual strength. During this time, RMTs continue their ongoing direct religious ministry that includes:

- Coordinating the availability of worship services, sacraments, rites, and services and/or ceremonies honoring the dead.
- Assisting with the integration of personnel replacements.
- Enabling the grief process through personal counseling and memorial services.

- Reinforcing the Service member's sense of self-worth and hope.
- Assisting with CEDs to provide opportunities for Service members to talk about their combat experiences and to facilitate integration of the combat experience into their lives.
- Providing leadership training and supervision of CEDs.
- Providing personal religious articles and materials.
- Providing or arranging faith group coverage in the command or for other units.
- A chaplain's ability to relate the spiritual aspects of life to the Service member's situation is an essential element in the replenishment process. Religious ministry can assist Service members in achieving emotional and spiritual wholeness.

Reintegration

After deployments and operations, RMTs can assist reintegration of the individual Service member into family relationships and society at large. Many religious ministry support programs remain the same. Expanded ministry may also include providing the following:

- Worship events for the entire unit.
- Worship events for varied faith groups.
- Briefings that help Service members recognize, prepare for, and master the stressors of reunion with family.
- Structured events to assist Service members returning to family and civilian life.
- Opportunities for Service members to experience and understand the forgiving and unchanging love of God.

Appendix C

Glossary

Section I. Acronyms

AAR .... after action review
BAS .... battalion aid station
BICEPS .... brevity, immediacy, centrality, expectancy, proximity, and simplicity
CED .... critical event debrief
CSC .... combat stress control
MCFTB .... Marine Corps Family Team Building
MCRP .... Marine Corps Reference Publication
MOPP .... mission-oriented protective posture
MRE .... meals, ready to eat
NCO .... noncommissioned officer
NGO .... nongovernmental organization
PVO .... private voluntary organization
RMT .... religious ministry team
RP .... religious program specialist
SGLI .... Servicemen's Group Life Insurance
SPRINT .... Special Psychiatric Rapid Intervention Team
UMT .... unit ministry team
WIA .... wounded in action

Section II. Definitions

circadian rhythm—Rhythmic variations in individuals; predictable physiological and behavioral cycle of about 24 hours.

combat stress—The mental, emotional, or physical tension, strain, or distress resulting from exposure to combat-related conditions. Combat stress is the net result of many stressors (e.g., fear of death, fear of failure, other intensive painful emotions [like grief], uncertainty, boredom, worry about what is happening back home, and the many physical and mental demands of combat duties) at any given time.

combat stress behaviors—The generic term that covers the entire range, from highly positive to totally negative behaviors which can be elicited by combat stress.

combat stress reaction—The common, predictable, negative, emotional and physical reactions of normal people to the "abnormally high" stressors of combat, which by definition interfere with mission performance or well-being, but which can be treated according to the BICEPS principles, with reassurance of normality and recovery, with rest, replenishment, and activities to restore confidence. The combat stress reactions do not include the misconduct stress behaviors, since the latter require disciplinary action.

continuous operations—military operations with many pulses of action every day and night, continuing for several days to weeks, which require careful planning and resource allocation to give everyone a minimum of 4 hours sleep in 24.

stress—An internal process of an individual for preparation in dealing with, or reacting to stressors. Stress involves physiological stress reflexes, which ready the body for fight or flight, such as increased release of adrenaline, heart rate, and blood flow to brain and muscles, with subsequent increase in alertness. Stress also involves the accompanying emotions (stress responses) such as anxiety, anger, or guilt. Stress may or may not involve conscious awareness of the threat, but the stressor must be perceived at some level of awareness. The amount of stress experienced depends on the appraisal of the stressor and its context by the individual, even if that perception is incorrect. The stress process includes psychological defenses which may shield the individual from perceiving more threat than he can tolerate.

stress behaviors—Stress behaviors are the observable behaviors that an individual exhibits as a result of internal stress. They may be intended to overcome and turn off the stressor, escape it, adapt to an inescapable stressor, or make the uncomfortable stress symptoms more tolerable. A stress behavior may simply be a side effect of the tension, which is part of stress. Any of these different types of stress reactions may be successful, unsuccessful, adaptive, or maladaptive. They may make the situation better, worse, or not influence it at all. They may resolve one stressor but create new ones.

stressor—Any event or situation that requires a non-routine adaptation or behavior change, creates conflicting motives within the individual, or is perceived by the individual as a threat to his well-being or self-esteem. Combat stressors are any that occur in the context of performing the combat mission (whether under fire or not).

sustained operations—Continuous operations or combat with opportunity for less than 4 hours sleep per 24 hours for significant personnel, which may be brief or fragmented.

Appendix D

References

DoD Directive

6490.5, *Combat Stress Control (CSC) Programs*

Army Field Manual (FM)

8-51, *Combat Stress Control in a Theater of Operations, Tactics, Techniques, and Procedures*

22-9, *Soldier Performance in Continuous Operations*

22-51, *Leaders' Manual for Combat Stress Control*

22-100, Army *Leadership*

90-22, *(Night) Multiservice Night and Adverse Weather Combat Operations*

Miscellaneous

U.S. Army Aeromedical Research Laboratory and U.S. Army Safety Center, *Leader's Guide to Crew Endurance*

http://www.mcu.usmc.mil/TbsNew/Coverpag.htm, *Human Factors in Combat: Topics in Combat Leadership that Help Preserve the Fighting Spirit and Prevent Psychiatric Casualties*

Made in the USA
Monee, IL
05 November 2026

67808331R00108